The unsolved puzzle
Interactions, not measurements

Jonathan Kerr

A new avenue of thought in quantum mechanics, which now has growing support, and removes the need to make observations important

Copyright © Jonathan Kerr 2002 - 2019

Published by Gordon Books
April Cottage, 9 The Green, Clophill, Bedfordshire MK45 4AD

The publication of this book establishes the author's origination of some of the ideas in it (and as intellectual property), as it precedes any other publication of them, unless by the author. This refers in particular to the theory in the later part of the book, which starts in Part 7, Chapter 31, that is, the set of ideas called dimensional quantum mechanics, DQM theory, or the dimensional interpretation for quantum mechanics.

26th February 2019

ISBN
Paperback: 978-0-9564222-6-2
E-book: 978-0-9564222-5-5

The right of Jonathan Kerr to be identified as the author of this work has been asserted by him in accordance with the Copyright, Designs and Patents Act 1988. All rights reserved. No part of this book may be reproduced or transmitted in any form or by any means, electronic or mechanical, including photocopying, recording, or by any information storage and retrieval system, without permission in writing from the copyright owner.

Publisher's preface

I first met Jonathan Kerr in 2002, when we were both working on a documentary about physics that was to be produced from France. Being a documentary maker and science writer, I was to direct it. Paul Davies, who I had a meeting with about the film, was to narrate and present it, and the producer told me of a British physicist, Jonathan Kerr, who would also make a major contribution. We both flew down to Nice for meetings. We liked each other while working on the film, but it was never made. The producer died before filming started, and although his father Louis Moriamé tried to take the project over, the rights were sold to an Australian TV company. Jonathan and I then talked about making a documentary about the ideas in the book he was writing at the time. He argued convincingly that the mystery of quantum mechanics would have a conceptual solution, and a purely physical one, but that we haven't found the concepts yet.

I told him that before I made the documentary, he should finish his book. I offered to publish it when he was ready. At the time he was investigating what was later called 'The Interactions Avenue', which was the name of the film I eventually made. He told me how experiments on quantum decoherence were showing that interactions cause the change from waves to particles, but that no-one knew why. He was always pointing out clues that had not been taken onboard. I felt he had a very good grip on the puzzle. For years he used to tell me he had half a solution, and was looking for the other half. He would sometimes show me what he was working on, but he showed it to more or less no-one else.

Since those days I have been his confidant, his friend, his publisher, and have enjoyed directing and presenting the documentary that has now been made. It was wonderful to work with Carlo Rovelli and Neil Turok. But when we started, I had no idea that Jonathan would take another fourteen years to work through the mystery! At times I became exasperated at the slow rate of progress. A lot of this book simply sets out a fascinating puzzle, more clearly than it has ever been done before. But his own solution, in the later part of the book, is devastatingly simple and lateral. It connects, very specifically, to a lot of pieces of the puzzle. Some of them have been unconnected until now.

Nigel Lesmoir-Gordon

With thanks to Nigel Lesmoir-Gordon for all kinds of help, and to his son Gabriel Lesmoir-Gordon, to Johnny Wood for his work on the artwork, and to Brady Senior for the dolphin painting. Also thanks for helpful email discussions to Ben Dribus and John Lucas, and to Carlo Rovelli and Neil Turok for long and helpful conversations, mostly on quantum mechanics, in Cassis and Cambridge during 2017. And thanks to Jill Pliskin, for all the days when we kept up mutual support, quite often through difficult times, during the work on the book.

The chapters in this book are short, averaging less than two pages. So the index is long, and is at the end of the book with the alphabetical index and reference section.

The Unsolved Puzzle
Book I

Introduction

This book starts by summing up the clues of a well-known mystery in physics: the one about what's underneath quantum mechanics. Then it shows how a new avenue of thought is changing the landscape. It's about a way to remove measurements, or observations made in experiments, from the picture. The idea now has support from some of the world's best known physicists.

Later on I'll set out a new solution, emerging from that angle on the puzzle. I won't stay on the effects and phenomena for very long, as the details can be found elsewhere. I'll just give an outline of some of the clues, including some recent ones, and some that are often overlooked.

Nowadays a lot of people have read about quantum mechanics. The pattern of events that happens at a small scale is well known, but not understood. At the heart of the puzzle, we know that somehow, for some reason, there's a sudden change to matter's state, and undefined waves turn into well-defined particles. This happens with both light and matter, and it's clear enough that something very fundamental is at work. So if we look for an explanation, or look for clues, we're trying to get at something universal.

The world at the particle scale is consistent. It goes by a neat set of rules with their own weird internal logic - we can predict exactly what happens. There are known events and behaviour, and to physicists, it looks very much as if something is 'going on'. So far, we don't know what it is, but whatever it is, it generates that weird set of rules naturally.

So the puzzle needs a particular kind of solution. Nowhere else in physics is there such a need for a *picture*, to slide in underneath what we observe, and underpin the mathematics that we've found to work so well. Such a picture would provide, or *be*, a good explanation. Some think no such solution exists, but puzzles sometimes seem that way until they're solved.

Some of the ideas in this book have also been put into a documentary for TV, 'The Interactions Avenue', which has conversations between myself and two of the world's best known physicists, Carlo Rovelli and Neil Turok. It's still a minority view, but they both think the change from waves to particles isn't set off by measurements at all, but by interactions between bits of matter.

Interactions between matter and matter, or light and matter, happen all the time at the particle scale, including in laboratories. To make a measurement, you have to make one thing bump into another, or cause an interaction. The technology has improved steadily for decades, and by the '90s an area called decoherence came into its own. And first theory, then experiment, started to show that interactions are making the change happen. But until now, no-one has been able to say why interactions should do that.

And the idea has questions and problems. But then so does every other idea in quantum mechanics. They're all that way, but this is a new direction, and a path worth exploring. Interactions do better than measurements in some key areas, and completely remove what some would call the unscientific aspects of quantum mechanics: mind, consciousness, the observer.

After going a little way along this avenue, I'll set out a simple, lateral solution for the puzzle. It's an unexpected one - few, if any, theories have interactions directly causing the change we observe, and this is the only one to provide a *reason* for why they should do that.

Recently in physics there has been a widening belief that our understanding needs to catch up with our mathematics, and that we need some underlying *conceptual picture*, to accompany the mathematical one. Einstein and John Wheeler both said they thought that such a basis for physics exists, and that it would be found in the future.

Einstein pointed out a good place to search for it: the question of whether matter has a definite **location** in quantum mechanics. In his later years, he wrote: "*...it depends precisely upon such considerations in which direction one believes one must look for the future conceptual basis of physics.*" He also said that Planck's early discovery of quantisation had "*set science a fresh task: that of finding a new conceptual basis for all of physics.*"

Wheeler said "*...it's not just a matter of nice, simple formulas, there's some ideas out there waiting to be discovered*". He also said at the bottom of it all, we'd find "*...not an equation, but an utterly simple idea*", and that when we find it, we'll ask ourselves "*how could it have been otherwise?*".

The solution set out here potentially fits those, though it may or may not be the conceptual solution they were talking about. But it's a lateral approach, coming out of a simple conceptual starting point with a visual picture, and it takes a very different angle from previous ones.

JMK

Part 1. The core of the mystery
1. Uncovering an unfixed world

In 1927 the top people in the physics world met at the 5th Solvay conference. They were there to discuss a new theory that had just been formulated the previous year: quantum mechanics. There's a well known photo of them - Schrödinger, Heisenberg, Bohr, Pauli, de Broglie, Dirac, Einstein, Planck and Born were all there. The new theory worked very well, and it fitted perfectly with experiments. They knew it was a major breakthrough, but the problem was, it was impossible to understand it.

What they'd found, which no-one could explain, was that when you make a measurement on matter, it jumps from one state to another - from waves to particles. Matter then takes on definite properties, and suddenly becomes more clearly defined. But this change seemed to happen only when someone made a measurement, which was very baffling.

From then on, people had the mathematics of quantum mechanics, but they had no explanation for what the mathematics was describing. We've been doing experiments on light and matter at a small scale for nearly a hundred years, and they always come out exactly as the theory predicts. But we still don't know what's going on.

Physicists don't like to say more than they *can* say. But this change was only observed when a measurement is made, so they had to bring the concept of measurement into the theory, and to some extent build the theory around it. It was hard to take an event separately from the measurement that sets it off. But they brought measurements in reluctantly, and for decades a lot of physicists were profoundly unhappy about it. It looked too like the observer might somehow have a special role - it was as if we were bringing the small-scale world into a more concrete existence by observing it.

During the 1930s, there were massive discussions worldwide about how the theory could be interpreted or understood. Then the discussion died down, and physicists rather stopped trying to understand it. Instead they just got on with learning to use quantum theory, and building the modern world. It was the most accurate and reliable theory ever: it passed every test, and led to the technology revolutions of the mid and late 20th century. The problem of interpreting it was left to the philosophers, who have been struggling with it ever since.

But in the 1990s the discussion started again, after new clues arrived. There had been very good progress on the theoretical and experimental fronts, and people started to see the whole process differently. Above all, an area called decoherence had given us a window into the so-called collapse of the wave function, the sudden change that happens.

We now know, coming from both theory and experiment, that the change isn't instantaneous. It takes a definable, non-zero, very short period of time. We can calculate it. Although decoherence leaves us with no explanation and still no real understanding, it had allowed us to peer further into whatever's going on. And what we see there looks like a rapid series of interactions.

One result of this is that the goalposts have moved, after staying in the same place for around seventy years. And an avenue of thought has emerged - the idea that interactions are what sets off the change - that looks like it might one day remove measurements from the picture completely.

2. An object made of possibilities

The particle scale, where quantum effects cause most of what happens, is a strange place. If we could see down to that scale, we'd see quantum waves spreading out in space, and when they do, matter is in a raw, undefined pre-state. Certain things about matter are undecided, for whatever reason. So the wave contains many overlapping possibilities: for things like what matter might be doing, where it might be located, even for what properties it might have. These possibilities are all superimposed - they all land on top of each other somehow, in nearly the same location.

The wave is like a little spread-out object at the particle scale, too small for the eye to see. It behaves like a wave, making interference patterns, as real physical waves do. So it seems to be a real object.

But somehow, it's also made of possibilities. When a measurement is made, one of the possibilities is picked out, and matter jumps into its well-defined particle state. It now behaves in a way we understand much better. It's as if a decision has somehow been made, and one possibility has been picked out from many.

For seventy years people struggled with the odd idea that observing matter is what sets off the unexplained change from waves to particles. It was as if the experimenter's mind was affecting the experiment, and bringing matter into a more concrete state of existence.

This led to all kinds of weird ideas, and not only coming from the fringes - the majority of mainstream physicists believed it too, and they still do. A physics Professor who conducted a groundbreaking experiment in 2015, said in an interview: *"It proves that measurement is everything. At the quantum level, reality does not exist if you are not looking at it."*

This kind of view quantum theory is very widespread nowadays, both among physicists, and among non-physicists who have read about it. But it's part of an unsolved puzzle, so it might or might not be true.

3. The cart and the horse

The reason we need to interpret quantum mechanics at all, and the reason there's a puzzle to solve, is that physics sometimes gets the cart before the horse. It's possible to find the mathematics of a theory before the underlying picture, instead of the other way round.

With the cart and the horse, it really matters which order you put them in. But with the mathematics and the conceptual picture in physics, the order we *discover* them in doesn't matter at all. Either is fine, but assuming they both exist, there's a need to get both.

When you can see the phenomenon, the picture comes first. With waves on the ocean, first we see them, and then we try to get to the mathematics that describes them and how they behave. But nowadays we're working on scales so small that we can't see anything much. We just do experiments, and that's how we get to the mathematics. So sometimes the mathematics comes first, and if so, you might be left with a puzzle: to find the picture.

4. Interpretations

The picture should lead to the mathematics. So if we have it, we can get the equations to come back out. But as some people may not realise, it doesn't necessarily work the other way round. The picture might not be *reachable* from the mathematics. Why should it be? Mathematics on its own could be describing a whole range of things, and it won't necessarily tell you which or what. This point might explain why ninety years of looking at the theory has left us with no consensus, and no real understanding.

We're very focussed on the mathematics, partly because that's all we have at present. But if we're going to find the picture, we might need to look away from the mathematics, and create new ways of making conceptual progress - we actually have very few systems in place for doing that. So that's what this book's about.

The mathematics of quantum theory has never been wrong. We don't know how to interpret it, but the different interpretations are different attempts at drawing pictures of what's going on. Some look more like visual pictures than others, but they all have an element of *explanation*, and the aim is to explain what's happening, whether or not in a visual way.

Although there's no consensus nowadays, Nils Bohr's view from the 1920s, the Copenhagen interpretation, which used to be way out ahead of all other interpretations, is still the standard view if there is one. It emphasises only what we can know, and dismisses what we can't - Bohr was influenced by positivism, which does the same.

The Copenhagen view was never defined in an exact way. Part of its role was to set out the early landscape. But its angle is that quantum theory doesn't show us a world that exists independently from us, but a world that we have to interact with to observe. As a result, what we can observe is limited. Bohr took the set of possibilities implied in the mathematics, that somehow exist before a measurement, to be about our potential *knowledge* of the setup, rather than something real in its own right.

Copenhagen doesn't necessarily mean 'no objective reality', as some take it to mean. It's nearer 'no objective reality that we can observe'. But it can be taken to contain the implied idea that we should hold off trying to interpret the theory. Bohr thought the underlying reality was unknowable, and his strong character and persuasive nature influenced generations of physicists to say 'that's just the way it is'.

It's worth mentioning that Bohr drew no distinction between measurements and other physical forms of interaction, so in a way Copenhagen included an early hint at the '90s interactions approach. In the 20th century Copenhagen was the main view we had, but there were four or five other approaches, some more marginal than others. Since then, other views have been catching up. There are now more than fifteen leading interpretations, and many more surrounding them. Each of them, including Copenhagen, has its own major problems. At times, trying to choose between them seems more like a choice between problems than advantages.

5. The wave function

The wave function is a mystery, if ever there was one. We simply don't know what it is. In quantum mechanics, what people call the wave function is a mathematical description of some kind of wave. It usually takes the form of one equation, the Schrödinger equation. The wave seems to be an unformed pre-state of matter, before certain details become well-defined.

Anything in the particle state seems to emerge out of these waves. The wave contains a set of probabilities about what will happen when it does, perhaps because the wave contains all the possibilities for when it does.

Before a measurement, there's a wave sitting there. It can make interference patterns, so it's a real wave of sorts, but it's also a set of possible versions of the situation. When the measurement is made, the wave gets prodded, and it immediately disappears. After that, just one configuration remains. So in a rather baffling way, it's as if a choice is being made - whatever really happens at that point, it's absolutely clear that it's more than just the measurement disturbing the system ('the system' is, loosely speaking, the chunk of matter that's being examined).

From then on, if the system is left as it is, every measurement gives the same result. Matter is now in its familiar particle state, which we understand well. We've known about matter in this state for a long time, and we assumed, naturally enough, that it was all that matter ever does. It was only a century ago that we caught it appearing out of a sea of possibilities.

A world that appears out of a sea of possibilities might look like it fits loosely with some of our other mental furniture. But physics is a discipline, and you don't make assumptions that you don't have to make.

The wave function might represent our potential *knowledge* of the system before a measurement, or it might be a real state of matter. Each of these two alternatives - information and reality - is a heading under which different interpretations are listed. But it has been hard to do away with either, and many think the waves are some mixture of the two.

In 2011, a poll was taken at a conference on quantum mechanics. The result showed a shockingly wide spread of opinions, and very little consensus. (A wide spread of opinions often translates to 'we don't know the answer'.) The Copenhagen interpretation was still out in front, but it had lost the large lead it had in the 20th century. And the idea of the wave function being a mixture of these two things got 33%, beating 'real' and 'informational', each of which on their own got 24% and 27% respectively.

These issues are about what's real, and because of that, we find there's a lot on the table. Since the 1920s, one of the main questions has been whether we can keep the notion of an objective reality. Matter seems to come into a far more clearly defined existence when we observe it, which has led many to let go of the idea of a reality independent of the observer.

But we need an objective reality to do science (it would be nice to have one anyway), and given that it's very much an unsolved puzzle, the answer might include one.

6. What *is* the wave function?

Over the last few years measurement techniques have been developed for observing the wave function more directly, which was always impossible before. If measurements are made very carefully, it's possible to keep the 'many possibilities' of the wavelike state, and avoid setting off the collapse of the wave into a single particle-like state. Experimenting near the place where this change happens, with what are known as 'weak measurements', and trying not to tip it over the edge, enables us to look for clues about exactly what sets it off.

It has long been thought that observations set it off somehow, but that idea has major problems, and nowadays it's beginning to conflict with what has been found. It turns out that it's possible to tip the wave function a little way over the edge towards collapse, and then tip it back again. These fascinating new experiments, pioneered by a Canadian team, have made us think hard again about what the wave function actually is.

And despite eighty years of progress, when people list the possibilities for what the wave function is, the mystery remains the same. We can compare two lists: In 1927, soon after the theoretical discovery, Born and Heisenberg were commissioned to write a report on quantum mechanics - the work was set in motion by Lorentz. A later summary of the report says "*...the idea that the state of a system is given by its wave function [...] is bound up with the question of whether the latter should be seen as a 'spread-out' entity, a 'guiding field', a 'statistical state', or something else.*"

Compare those possibilities with a recent list. In a 2009 article in Scientific American Magazine, David Albert and Rivka Galchen wrote about the wave function: "*But <u>what</u> is it exactly? Investigators of the foundations of physics are now vigorously debating that question. Is the wave function a concrete physical object? Or is it something like a law of motion or an internal property of particles or a relation between spatial points? Or is it merely our current information about the particles? Or something else?*".

It shows how broad the spread of possibilities still is. Over the eighty years between these two lists, the ideas we have for what the wave function might be still cover a very wide range of different *kinds* of possibilities. We haven't narrowed it down much. And the list still includes, tacked on at the end, 'or something else'.

The fact that we haven't narrowed it down, arguably, might *suggest* that it's something else. Later on in the book I'll describe a picture for what the wave function is - it's entirely new, and it definitely comes under 'or something else'. It has long been thought that the wave function represents the state of our knowledge of a system, but strong arguments have been put forward recently that it somehow represents reality.

In 2011 three physicists, from Imperial College and London University, found a theorem that has had a major effect in the quantum mechanics world, and is seen as the most important breakthrough for perhaps fifty years. Pusey, Barrett and Rudolph immediately became known as PBR (which echoed the nickname EPR, for the 1935 paper by Einstein, Podolski and Rosen). What they did, like John Bell's theorem from the 1960s, limited the possibilities about how quantum mechanics can be interpreted.

PBR has made it more likely that the wave function is real, and less likely that it's epistemic, that is, about our knowledge of the system. More recent work has pointed the same way, going via a different route. PBR has shown again that mathematics can reduce the possibilities in important ways, but our attempts to understand the wave function is still unavoidably an area where some kind of *conceptual* shift is needed.

Shan Gao, also in a 2011 paper (whether or not he was influenced by PBR), looks at the vital question of the conceptual nature of the wave function. He takes it to be real, and says that being real, it must be one of two things. The first is a physical field, but he rules that out because he says if so, there'd be self-interaction. The other possibility, he says, is that the wave function is the ergodic motion of a particle, which is the discontinuous and partly random motion of a single particle, and he concludes that it's that. (This is actually similar to Rovelli's view, which is that the motion of matter is discontinuous, and you can't fill in the gaps.) The suggestion I'll make later on for the wave function is a third possibility, not like either mentioned by Shan Gao, and not like any other existing view of it.

7. Quantum eraser experiments

Some of the most surprising discoveries with quantum mechanics have been made in 'quantum eraser' type experiments. They started in 1982, but recent versions have been increasingly revealing and surprising.

Physicists send photons around a circuit, and then play cat and mouse with the information available in the experiment: turning a blind eye to it, peeking at it - afterwards, upside down, blindfolded, through a computer - always knowing that if they look a little too closely at which route the photons took, or create 'information' that's a little too specific, they'll find the behaviour of particles, while if they look less closely, and ask less specific questions, they'll find the light behaves like waves.

But it's not observations doing it. They create 'routing information' (using the phrase loosely), that sets off the change from waves to particles, then erase it, to see if the system returns to its previous state. And it does. It's as if the mere existence of the routing information leads to particles, but no routing information leads to waves. The key point is: we don't have to look at the routing information to switch between them.

Other experiments are closing in on other aspects of it. It's an exciting time, as our technology has finally got good enough to probe the quantum scale freely. Quantum entanglement is a prediction from the theory that says pairs of particles can be linked across space in a very direct way. Making a change to one should affect the other instantly. We know of no physics that could allow this, and it seems to compromise the speed limit c from relativity. But we also know that quantum theory has always been right so far.

To confirm that these correlations exist, people have had to narrow things down in the experiments, carefully making sure that it isn't something else. In 2015 three separate teams finally managed to perform entirely 'loophole free' experiments, creating these links between particles, and strongly ruling out connections at lightspeed or slower.

One thing this does is to rule out the idea that any *signal* passes between the photons of a kind we understand. But until we know more, we can't say if the speed limit is being compromised. It seems that there may be something unknown, for which c simply isn't the speed limit. If so, because its 'effect' is instantaneous, it isn't really travelling at all.

In other areas of physics, we want our sets of rules to be universal, even if the evidence says that they're not. Some try to apply the rules for matter to light, saying things like 'because it travels at c, a photon doesn't experience time'. But the fact is, we sometimes find that light doesn't go by these rules for matter. Many good physicists know that applying them to light means depending on an unreliable assumption. And with entanglement, some try to apply the travelling rules for *both* light and matter to something else. But whatever it is, it clearly doesn't go by them.

It's sometimes claimed that quantum eraser experiments show observation to be all important. In fact it was a brilliant experiment because it managed to probe what happens enough to find out that observation is irrelevant. It led to an absolutely vital bit of knowledge that took us eighty years to get to:

that we don't have to look at the routing information. Instead we can create it, then erase it without looking at it, and somehow its very existence, or lack of it, seems to switch between waves and particles.

Some say the fact that we *could* have looked at the information is enough. But why 'we'? Why assume that we're involved? That comes merely from past habit. We've had decades of assuming that the observer is central, and it certainly used to seem so. But the fact is, the 'we' no longer applies. The experiment could be done by robots a hundred million years after humans become extinct, to put it in (rather bleak) perspective.

So what could the cause be? Well, to create the quantum eraser experiment as it is now, we've assumed entanglement exists. It's not so relevant that it's used in the experiment, the point is, we believe it exists - even though we don't understand it, and even though entanglement behaves exactly as if there's a 'holistic' system at work, which is somehow connected up instantly, and is capable of working as a whole, across distances in space.

That could be a glimpse of part of the cause. So if we're wondering about this routing information (or lack of it), that's able to switch between waves and particles, and *who or what it affects* - 'the experimental setup' is a far better candidate than the experimenter.

The interactions approach, which we'll get to next, isn't enough on its own to explain quantum eraser experiments. But suppose something is working in tandem with entanglement type effects. If so, and it seems so, then because we now know observations aren't needed, interactions looks far more likely than consciousness. So *interactions and entanglement*: perhaps these are the raw ingredients.

And as we'll see later, it turns out that they're closely linked: entanglements are created by interactions. We can guess in this kind of way, and these days what we're looking at looks more like a physical system than it ever has. But we can't infer much about what a correct interpretation would look like. For now, the point about these experiments is that although they certainly do weird things, they're all consistent with quantum theory. As often happens, yet again, our theory tells us exactly what we can expect, but it doesn't tell us how to interpret it. Or to put it another way, we have the mathematics, but we don't know what's going on.

N.B. An important experiment was done recently which supports the ideas in this book. The paper was published in February 2019, just before the book was due to come out - there's an additional note about it on page 106.

Part 2. The first signs of a change

8. A new avenue of thought

The basic puzzle of quantum theory hardly changed for seventy years, but by the 1990s a new idea had appeared. It was incomplete, without a doubt. But it looked like it might eventually remove the 'measurement problem', and leave the mind of the experimenter out of the process completely.

One of the pioneers of the idea was Carlo Rovelli, who developed a new way of seeing quantum mechanics, which brought it more into line with special relativity. His interpretation is called relational quantum mechanics, or RQM. In a relational theory, a lot of things depend on the viewpoint that an object is seen from. In making quantum mechanics a relational theory, Rovelli had to shift a lot of concepts around. And in the process, he looked behind the measurements, to see what was underneath them.

And what he found behind the measurements was something we'd always taken for granted before. But it was something that others were now starting to focus on as well, because of decoherence: interactions between bits of matter. Decoherence had already shown that somehow, interactions were what causes the wave to disappear.

But no-one understood that, so apart from in Rovelli's work, there was very little mention of them - that is, outside decoherence. But Rovelli, although he said that he had no explanation for it, got to the idea that the change from waves to particles is not caused by measurements, but by interactions. The point is, when you measure something, you have to make something else interact with it.

So perhaps when we make a measurement, although we think that it affects things in some other way, in fact we're just bumping bits of matter together. And perhaps this bumping together of bits of matter does something extra that we didn't know about, and that we simply don't understand yet.

By the mid '90s, RQM and decoherence were suggesting that interactions are important. And by the late '90s, even experimental results - coming out of decoherence - were confirming that interactions play exactly the role that the theory suggests. This was a real change of direction, and right where one was needed: away from measurement, away from the observer. But no-one knew what it meant.

(Some have argued that the *lack* of a measurement can set off the sudden change, if we prepare the experiment in a particular way. But as in the first chapter of Part 10, given that we have no understanding of either concept in relation to this, it could equally well be the lack of an interaction.)

So what could interactions be doing that we don't know about? That's the question. It's a comparatively new question, and it looks like a difficult one. We know they'd have to be doing a lot more than just disturbing the matter that's being examined. Instead, an interaction would have to do *something* that changes matter's state completely. So even though we now have a way to throw out measurements, if we're going to explain this, we're still going to need something that changes matter's state.

So far, no-one has been able to suggest a reason for why interactions should do that. Partly because of that, the avenue of thought about interactions has been something of a taboo, and has not travelled far. It's not that physicists are reluctant to talk about it, far from it. It's more a case of - no-one from outside their area asks them about it, because no-one from outside their area knows about it.

Anyway, although interactions is a minority view, that minority is growing, and it includes some of the world's top physicists. Nowadays some very good physicists believe that measurements are nothing to do with it at all. So the exciting thing is, we now have a new question to answer. The goalposts have moved, so now we might solve the puzzle.

9. Decoherence

These days decoherence is seen as having replaced the idea of the collapse of the wave function. It's seen by many as an essential part of the current view of quantum mechanics. Decoherence provides a more detailed account of state reduction, which means, loosely speaking, the change from waves to particles. And although we still have no understanding of the process, we've found out a lot more about what actually happens.

Decoherence is not a theory that accompanies quantum mechanics, it's part of the original theory. What was found came out of the basic mathematics we had already, and is a direct consequence of it. It took decades to work through the consequences, but if one looks closely at what quantum theory says about a situation with matter in the wave state, surrounded by other matter in a laboratory, you find that the original theory predicts more than we thought.

The process wasn't found sooner because we were used to classical physics. And in classical physics, it's no problem to take the object being examined to be isolated from the rest of the world. Other objects some distance away can be irrelevant, if we choose to make them so. But at the particle scale, it's not like that. If you put a new particle in a place where there are other particles around it, something that we don't understand happens.

In decoherence, a chunk of matter's interaction with the surrounding matter affects it in a way that erodes the wave function. If the wave isn't shielded from its environment, it will interact with its nearby surroundings, rapidly creating entanglements with other bits of matter. And as it does this, the wave state (made of many superimposed possibilities) loses its coherence, or decoheres. That means the number of possibilities it contains is reduced, and it sometimes goes down from many possibilities to one.

That's a loose description of the existing picture, the road to which was long, more than half a century. The roots of it go back to Everett, Bohm, and more recent contributions from Zeh, Zurek, Schlosshauer and others. Zurek's work is central: he worked away almost unknown in the '80s, but was much better known by the '90s. The present understanding of decoherence started with a 1991 article by him in *Physics Today*.

Some things have changed, others have stayed the same. Just as it was in the past, when the wave disappears, there's a lot we don't know. We don't know what the possibilities are that are getting removed, we don't know how one of them is picked out, we don't even know why the process *does that*. But the good news is, we know a lot more about what happens when it does.

And what happens is a rapid series of interactions, creating entanglements. It takes a definable period of time. Zurek published an equation for deriving this period of time as early as 1984, and many years later it was supported by experiment. And the larger the object, the faster it decoheres - we now know that from experiment as well. It has shown something about the relationship between the large-scale world and the small-scale one, which in the past was even more of a mystery than it is now.

These interactions occur routinely without a measurement, and they bring sudden changes. But because a measurement *also* involves an interaction, it now looks like in that instance as well, when a measurement is made, and a sudden change happens, it's just the interaction that's doing it.

But if it's just the interaction that's doing it, then it all starts to look different. It starts to look as if this process, whatever it is, may happen with or without a measurement, with or without an observer, and therefore with or without an observer's consciousness. So it starts to look like that whole approach, involving measurements, observers and consciousness, may have been a bit of a red herring.

10. Three stepping stones

So perhaps this change in our view is about landing on the second of three stepping stones. There are three steps that show where our understanding might go, over time. The three lily pads on the back cover of this book, across which a frog might hop, can vaguely symbolise them.

The first step is taking quantum theory in terms of measurements, as we did in the 20th century. The second step is taking it in terms of interactions, but without understanding that, which is more an early 21st century view. The third step is going on to understand whatever it is that interactions do. And I hope it doesn't take until the 22nd century to get there.

Part 3. Ideas and clues

11. Picking out the clues

Somewhere among all the things we know about quantum mechanics, there will be a bit of information that could tell us something. In physics we get a long list of facts and information, but only a very small number of those will be clues that we can use.

A clue is often a connection between two things. It might be an analogy, that shows one thing is similar to another. If we find a link between something in quantum mechanics that we don't understand, and something else that we do understand, that might help. But even a link between two things we don't understand can help. Sometimes you can dig away at the edges of that, and find out more than we knew before.

At the end of this section, in Chapter 18, I'll set out a major clue that would have helped if I'd seen it. I only saw it later, with hindsight, and at that point I was very surprised not to have seen it before. This idea involves connecting two things that have until now been taken separately. Whatever one might think of the idea itself, it's odd that it has never been pointed out (as far as I know). But there are many examples of people's *expectation* affecting what they see. That's why I missed it, and here's a more down to earth example of the same thing.

In Cornwall earlier this year, I was looking out to sea from a coast path, and saw a huge grey bird with curved wings, flying over the water a little way out from the shore. It was bigger than a cormorant, dark grey and very different, flying several feet higher above the water than a cormorant would. I didn't know what kind of bird it was, but assumed that although not an albatross, it was some large seabird of that general kind.

Later that day we met some bird watchers on the coast path, and got talking with them. They told me that it might have been a heron, and I realised that it could have been. You see them a lot in Surrey, by the rivers there. It looked a little different from above, but that idea suddenly fitted with pictures of herons I have in my memory.

The point is, expectation affected what I saw. I didn't expect to see a heron, because herons are filed, I suppose, under non-seabirds, or freshwater birds. You expect seabirds over the sea. So the box I had to 'think out of' was the filing system, which can stop you seeing what's in front of you.

In the same sort of way, expectation will limit what we're capable of seeing, when looking at any puzzle. It will limit the connections we make, and it may be for that sort of reason that the clue pointed out in Chapter 18 was missed, certainly by me. With hindsight it jumped out a mile, but not before.

And yet with both that clue and the heron, everything that was needed to make the connection was already there. So it could have been made. All I can say is - and it's easy to say it - if we're looking at something and we don't know what it is, we have to try to throw out preconceptions, and not make assumptions about what we're going to see.

I can't say I managed to beat this kind of obstacle in any direct way, instead I got to an answer for quantum mechanics via a far more roundabout route. Lateral jumps are looked at in Part 5, and some different kinds. But one kind (which was missing that day on the coast), is a jump across headings, and to a different part of ones filing system.

12. The man who works in the laboratory - should he be removed?

Although we don't understand the foundations of the quantum world, we've been experimenting on it for ninety years, and we've learned more than just a few details. We know the general aspects of the world at that scale. And a close look shows that we can't ignore the gap in our understanding, and say 'that's just the way it is'. There are things that unavoidably need explaining - the role of the observer is one of them.

In quantum mechanics, the role of the experimenter has been enormously difficult to remove. And yet it must be removed if science is to hang onto its definition of reality. On the face of it, without observers to bring matter into a clearly defined state by making measurements, the world would always be in some unformed state, and only conscious creatures like us would be able to bring it into sharper focus.

Eugene Wigner gave the most respectable version of this view, but the idea has been widespread for a century among physicists and philosophers. And there are also others who take a less rational and scientific approach, who are drawn to the idea that mind, or consciousness, affects the experiment. They tend to ignore the fact that it's an unanswered question, and talk as if there's only one possible kind of explanation. I realised after a long time that in my view, the unhelpful assumption that both groups make is that we have all the pieces of the jigsaw in front of us. During the 20th century, science had been so successful that people started making that assumption, and it held us back in many areas. There was good reason to think pieces of the puzzle were missing. But if one assumes we have them all, one might make other false assumptions, about how the pieces can and can't be arranged.

For the first fifty years, the public weren't told much about the mysteries of quantum mechanics. What had been found was so weird that few physicists felt like describing it in simple language for the public. Then in the '70s Fritjof Capra, and in the '80s Danah Zohar (who talked about 'quantum psychology') and others, wrote books trying to link the theory loosely with 'metaphysical' or theological ideas. These kind of books are widely read nowadays, but they often give a false impression of where we stand with these questions.

There was also the 2004 film *'What the bleep do we know?'*, which to me is very misleading. Ironically it sounds, from the title, like it must surely leave a bit of room for these mysteries to be unsolved, but it doesn't really. Instead it suggests rather the opposite, and implies that we know enough to back up some ideas from outside science using quantum mechanics. The main theme is that reality is all in the mind - I'd say you can believe that if you like, but you can't claim it follows from quantum mechanics. The film, and the sequel, wrongly imply we're in a position to draw certain conclusions from quantum theory. It talks as if we know what's going on underneath the experiments, with lines like 'the conclusion is inescapable'. And yet these films were made a decade after physics from the '90s had shown that various conclusions were not inescapable.

The truth is, as many good physicists know, that until something is explained, it isn't. So we simply don't know what's going on. That means we can only go by what comes directly from the theory itself, and from experiment, such as the apparent randomness. We can't depend on, or draw conclusions from, what comes from any assumed interpretation.

Anyway, among physicists who neither avoid the issues nor draw premature conclusions from them, the hope is that the observer will be removed, and that physics can keep mind and matter separate, in the wider world outside living systems anyway. For one thing, the universe seems to have been here before we were, and before there were conscious creatures hanging around in it. So if consciousness brings the world into existence, as many believe, there are questions about how the world got to where it is now.

To deal with that, some well-respected physicists have even taken the view that our consciousness bring the world into existence in the first place, acting backwardly in time to cause the big bang, and creating a time loop that goes round and round. This is one of the 'desperate remedies' for the problem that have been suggested (the phrase was first coined by John Gribbin in the context of quantum theory, and is the name of a Thomas Hardy novel). The weirdness of this idea, along with its enormous lack of confirmability, shows just how difficult to deal with the puzzle has been.

But not many physicists want consciousness to be in the equation. Hopefully physics won't go off into weird untraceable stuff from the point we've now reached. It would be good if the clear, rational systems we've developed for studying the world, and pinning down what happens in it, go on being useful in the way that they have been. So one of the hidden questions, when we ask if there's an objective reality out there, is about the future of science. The fact is, if science is to go on being the good thing that it is, we're going to need one. And for those who think science is a gift, there's still the possibility that it may be a lasting one.

At present, one way to remove the person who works in the laboratory is to make an unspecific statement, and leave holes in the jigsaw around it. Better to do that than to try to complete the puzzle before we have all the pieces. We can say: 'light or matter changes its behaviour from waves to particles *if information of a certain kind comes into existence.*' (The word 'information' is not necessarily being used as it's sometimes used - but other words, such as 'specification' or 'knowledge', are worse.) So if we create routing information about the setup, by putting photon detectors into the experiment, the light behaves as particles. This may apply if that information exists at all.

It sometimes helps to take a step backwards, to where the road divides into many roads, and admit that we're at a place with various avenues leading off from it. Saying only what we *know* we can say makes it less likely that we're already walking down the wrong road. I've found it helpful to take this kind of approach in many places. If we say the change happens if information of a certain kind is created, the new question now simply asks why creating that kind of information has that effect.

13. The double slit experiment

Richard Feynman used to say that the double slit experiment contains all the unexplained aspects of quantum theory rolled into one experiment. This was a deliberate oversimplification, but it does go to the heart of the matter, and other questions often reduce to this one. The experiment has been covered a lot, but here's an outline.

Sending light through two slits in a barrier gives it a good chance to show its dual nature. Matter does the same, but at this point it's easier to talk about light, just as in other areas it's easier to talk about matter.

The light can behave exactly like waves and pass through both slits, creating an interference pattern on the other side, as waves do. Or it can behave like a stream of particles, each of which goes through one slit and not the other. That creates a different pattern on the screen behind. It's easy enough to tell which is which - one pattern means waves, the other means particles. These two kinds of behaviour, and the resulting patterns, are very different. And if we do certain things, we get sudden jumps from one pattern to the other.

But at first glance, this dual nature of light can look like something else, and not weird at all. Water makes waves, but at a small scale it's just molecules. Water has a wave nature and also a particle nature. The wavelike behaviour is emergent, and appears at a larger scale. One might think the ambiguous behaviour in the double slit experiment is simply that kind of wave-particle duality.

But it's much weirder than that. Ralph Baierlein put it clearly - he said *'light travels as a wave, but departs and arrives as a particle'*. We get precise start and end points, but nothing specific on the way. John Wheeler said the same thing in more colourful language - he talked about a 'great smoky dragon', with the tip of its nose and tail visible, but in between them we can only see smoke - that is, murky, unclear, undecided states.

The pattern on the screen at the back tells us which aspect of the light we're seeing. The experimenter chooses which aspect to bring out of the situation, by choosing whether or not to put detectors in the photons' path, so finding out which route they took. And if you trace the photons' path exactly, they behave like particles - but if you don't ask too many questions about where they went, they behave like waves.

At the end of the light's journey, it hits the screen at the back. Then it seems to change from waves to particles anyway. There are many different versions of this experiment, including 'delayed choice' ones, where the light seems to decide how it travelled through the experimental setup after it gets to the other side.

It's natural to try sending single particles through the experiment. If the light source is adjusted way down, so it sends one photon at a time, then surely the behaviour of the photons should be separate and unconnected. Each one is detected at a particular point on the screen when it arrives.

But if that setup is left to run, the marks on the screen at the back where the photons arrive will build up into a pattern. And as more dots arrive, say, one every hour, what appears is the interference pattern again, the trademark of waves. This is shockingly difficult to explain, and the more you look at it, the more you realise what a conundrum it is.

14. The parts and the whole

It's as if the photons 'know' where the wave that they're a part of should go. And even though the wave is only hinted at in a statistical pattern, it's still as if the wave links them up somehow. And even worse, it looks as though each photon went through both slits, because if they each went though just one, you'd get a different pattern on the screen (and when they do, you do). But if they somehow go through *both*, then it looks like the wave is made up of particles in an unexpected way. Even taken separately, on their own, they're still connected to the wave - or to where the wave would be if there was one - and behave as if they're a part of it.

So it's not like the water molecules. With water, if you zoom in on it, and get down to the molecule scale, the water has lost all its wavelike characteristics. Sure, the molecules may roll and sloosh around a bit, but pour them through two plugholes, and they'll each go through either one or the other. But with light or matter at the quantum level, the wave seems to be so fundamental that even if you break it up into its components, they still behave like little parts of a big wave, and they each seem to have the *global* properties of the wave still with them.

When the one-photon-at-a-time experiment is done, the pattern that builds up on the screen is random in a way, but not in another. Where each particle lands is random within a particular area, but it still has probabilities attached to it. So although the pattern appears in a haphazard way, it's not entirely random. It takes the shape of a pattern that waves make, but that particles don't, or shouldn't according to intuition. And the pattern the photons make seems to exist across space, perhaps across time, and across whatever we put in its way to break it up into separate components.

It's as if there's some template underneath there somehow, perhaps not a guiding wave, but just some kind of unified existing thing, that is made up of many individuals, but which is also a single thing. In the mid 1980s, looking at quantum mechanics for the first time, thirty years before finding a solution to the puzzle, I briefly wondered if this ambiguity between the unified whole, and the individuals that make it up, was about some underlying principle of nature.

One finds apparently similar things in biology, where a group of cells behave either as individuals, or as a single whole creature. Perhaps the wave is like a flock of birds. Somehow each individual bird has a connection to the flock, so the flock can seem to behave as a single thing, and more than the sum of its parts. At the time, it seemed possible that the tension between reductionism and holism in physics was connected to this odd relationship between the parts and the whole. But this was only a loose idea, and it led nowhere. With hindsight, it turned out not to apply to quantum mechanics at all.

To me, it shows an interesting sounding idea can simply be wrong. But I've found this many times. It also shows how the quantum mechanics puzzle can lead us off into loose thinking, and away from scientific thinking. In my view, then and now, scientific thinking has been where we need to stay. That's the only place where things can be pinned down in a precise way, so it's the only place where we can actually find anything out.

Of the few ideas I had in the '80s, that one, about the relationship between the parts and the whole, was after reading David Bohm's book 'Wholeness and the implicate order', about the pilot wave interpretation. That was years before finding a solution that worked, the first glimpse of which appeared in the late '90s.

15. Information and reality

But then and now, in quantum theory the need is for analogies. The only way to crack into this may be via conceptual thinking, which means analogies and visual pictures. That's what I felt was needed, and I started to see the general area as a side of physics that had too often been overlooked.

The universe is a place that provides analogies. That's simply because there are a lot of things in the universe that are like other things. So hopefully the analogies are out there that will help solve the quantum mechanics puzzle. If not, there's really no puzzle, as there wouldn't be any solution. Instead, we'd just have the mathematics for what the world does, but we'd never be able to visualise what happens, or explain it. Explaining something means greatly reducing or simplifying the description of it, via an underlying layer of action. And it should be in a way that 'clicks into place' - that we can latch onto and understand. Visual pictures are often very good for explaining things, firstly because they're economical, and say a lot from a little, and secondly because our way of understanding things uses them.

Some of the people who think there's no interpretation for quantum theory, (and some go by mottos like 'shut up and calculate'), may simply think there's no analogy to be found that's comprehensible to us. It's a reasonable enough view, but the universe contains a lot of analogies between its things.

Anyway, pictures and analogies is what I was casting around for at the time, but they were few and far between. By the early '90s I had found a far-from-important analogy, which came after reading one that Paul Davies found. His analogy wasn't necessarily a very good one, and he said so. He mentioned the need for a better one, and with that I was off, and started searching. But the phrase he used alongside it was interesting at the time - he talked about 'the software-hardware entanglement'.

Now that, like the omelette analogy in Chapter 29 which it resembles, helped me to get some kind of a grip on things, without really understanding them. To some the phrase talks about mind and matter, but to me it was just about information (software), and real existing objects (hardware), being mixed up together somehow in the nature of the wave function. First, it behaves like a real physical wave, making interference patterns, as many kinds of waves do. But then it also behaves like a mathematical space, with probabilities pasted all over it, about what might be happening at each point. So it's a real wave and an informational one at the same time.

I don't know if he still thinks the same nowadays, but at the time Paul Davies thought the software and hardware were affecting each other. The analogy he had was of a computer with a robot arm, that reaches into a hole in itself, fishing around in its own circuitry. It then makes changes, affecting what the arm does next, in a feedback loop.

When I found an analogy for what the software-hardware entanglement was to me, it was about the relationship between the map and the territory. One describes the other, and the two are normally separate, and different sizes. But suppose the map is at a scale of 1:1, and drawn onto the territory. Every hill and bump in a landscape now has its description superimposed. And then one day you don't even bother to draw it on, which is a lot of hard work. You just use the territory itself as the map.

All this did really was to help bring that general clue into focus - the clue was that *something* was doubling as both reality and as a description of reality. Many years later, this turned out to be right, within my way of seeing things. It didn't necessarily help with getting to a solution, not consciously anyway. But when I did get to one, it helped with recognising it.

Just as an aside, looking at the nature of puzzles generally, I'd say that one of the marks of a good puzzle is how many solutions it *rules out*. That is, if we're looking for a solution that gives a clear, specific explanation, rather than just a vague set of ideas. The thing about the quantum mechanics puzzle is how many ideas it rules out before you start. The double slit experiment shows it well - it's the way that it seems so difficult to guess what's going on, because of all the things that it just couldn't possibly be.

16. Some different ways of seeing it

Before light or matter chooses a particular state, there's a set of possibilities about what might be happening. One way to explain this is to say that each possibility becomes reality, but in a separate world. These parallel worlds all differ from each other, at least in some very small detail. The many worlds interpretation began with the work of Hugh Everett in the 1950s, and many now have it as their preferred view of things.

One problem that many worlds has is that it can't explain the probabilities. What's known as the Born rule doesn't come into Everett's interpretation, but is on a list of things that are effectively thrown out, which are not seen as fundamental. According to Matt Leifer, some current versions nevertheless use equations that came out of the probabilites.

Anyway, standing further back, the many worlds interpretation comes under one of two headings, as most interpretations land in two basic groups. There are two ways of seeing the choice that seems to be made at the collapse of the wave function, and they differ about what exists *after* it has been made. In the first, after the change, only one possibility remains. If you simply go by what we observe, it seems that only one version of things exists afterwards - the one that is selected. All the others were possibilities, but somehow one of them is actually promoted to real existence.

But in the second kind of view, they all still exist somewhere. That might be in parallel universes, or separate histories, or branching threads that co-exist somehow. People develop these kind of ideas partly because a selection that entirely removes all the other possibilities is hard to interpret.

And yet the possibilities in the wave function are about what might happen in this world. They seem to have probabilities attached, as both theory and experiment have told us. Some are more likely to happen (in this world) than others. That all looks rather this-world-focussed to me, as if these variations in likelihood might somehow arise from conditions in this world.

If so, to find an analogy that fits this concept, we might need to find a choice that removes all the other possibilities completely. We don't know what the choice is that's being made, but we know that the different results have odds attached to them. And as far as we know, only one remains afterwards. So if they're all in this world, the other possibilities may all vanish.

Now if this is about what we can *know*, and information, then there you see one part of an explanation. What promotes one possibility to reality is our knowledge of it, so the others all disappear for that same reason. But there seems to be a physical side to the wave, and if so, the fact that the other possibilities disappear would be an important clue about what kind of choice is being made.

When we work out the odds for this choice, they make interesting patterns. The image on the next page (thanks to Kyle Forinash for the use of it), shows the probabilities for where an electron in a hydrogen atom might be. All the locations mapped out in the image are places where the electron might be found, and they're all connected mathematically for some reason. But before a measurement is made, the electron has no location, and it isn't in any one place unless or until it's found. Finding it fixes its location. Before that, there are places represented by the lighter areas in the image (all of them here in this world), where it's more likely to be found.

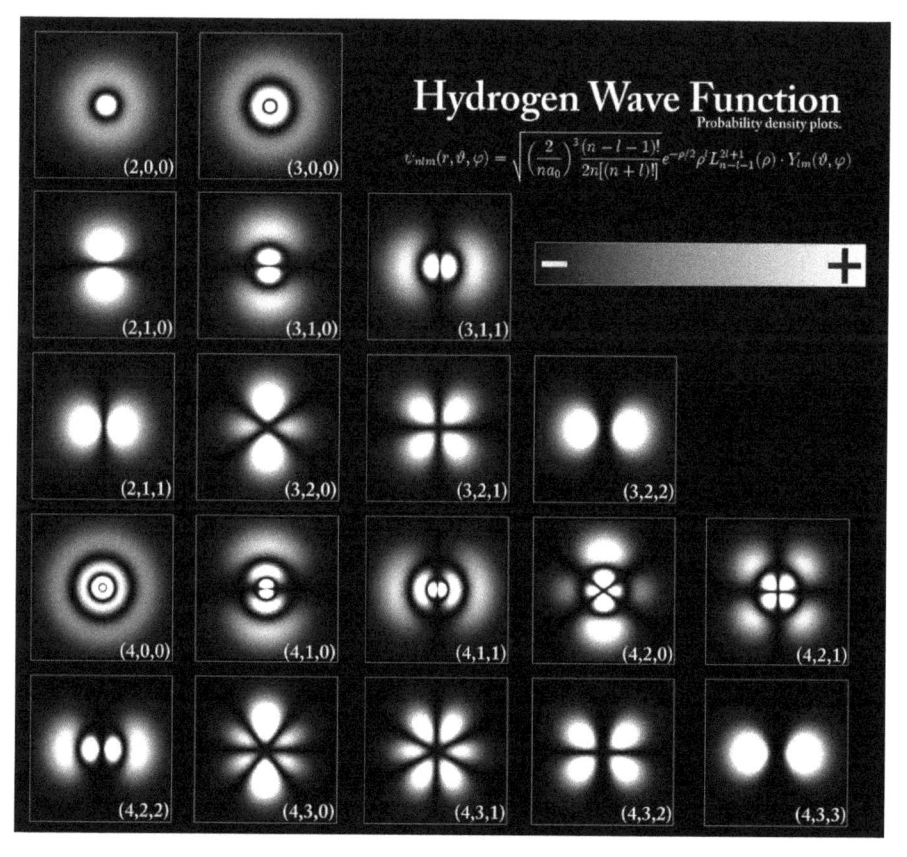

Hydrogen Wave Function
Probability density plots.

$$\psi_{nlm}(r,\vartheta,\varphi) = \sqrt{\left(\frac{2}{na_0}\right)^3 \frac{(n-l-1)!}{2n[(n+l)!]}} e^{-\rho/2} \rho^l L_{n-l-1}^{2l+1}(\rho) \cdot Y_{lm}(\vartheta,\varphi)$$

(2,0,0)　(3,0,0)
(2,1,0)　(3,1,0)　(3,1,1)
(2,1,1)　(3,2,0)　(3,2,1)　(3,2,2)
(4,0,0)　(4,1,0)　(4,1,1)　(4,2,0)　(4,2,1)
(4,2,2)　(4,3,0)　(4,3,1)　(4,3,2)　(4,3,3)

What could these patterns be...?

The regularity of these patterns implies something with symmetrical physical properties, and suggests there might be something almost mechanistic going on. And yet there's an element of chance involved. This mixture of precision and undecidedness is strange to us, but in fact, probabilities with an element of randomness do make patterns.

But that's different in certain ways. In the large-scale world, there are plenty of processes that have an element of randomness, and lead to symmetrical patterns. You could stand there throwing a tennis ball on a string attached to a pole, recording the results with a camera. You'll never get funding for this, but after many throws, you might find there are patterns, and perhaps even a bit like the ones above. But that isn't like the wave function. With enough information, the randomness will tend to go away, and it becomes possible to predict what will happen.

What we find down at the particle scale might look similar, but instead we've come to a non-negotiable barrier. The randomness there is no longer caused by approximations in the calculations, due to lack of information. Instead it's fundamental, and it doesn't go away. That's hard to explain.

Anyway, there's a range of approaches to quantum mechanics, and recently one area has become increasingly important. Quantum information theory is a comparatively new field, and has opened up a whole new landscape within quantum theory. Taking the theory in terms of information has led to a new set of rules, which might be capable of replacing the old, familiar set of rules. Attempts have been made to 'reconstruct' quantum theory from this angle.

Now when two different approaches lead to the same phenomenon, either might be more fundamental. That is, either might be more closely connected to 'what's going on', if something *is* going on, and more revealing about it. Both approaches provide a set of simple but counterintuitive rules, which the particle world seems to go by. Nowadays information theory is a major part of attempts to move forward in quantum theory, and some think that we'll eventually replace everything with information-related concepts.

But the older approach and this newer one are both ultimately mathematical approaches, and short of conceptual comprehensibility. The earlier approach uses familiar ideas, like waves, particles, superposition, entanglement - but we still can't understand them. The information approach is less familiar, but in both cases we just have to accept an unexplained set of rules.

In quantum information generally, people tend to talk about measurements, not interactions. There are different interpretations within it, but in a way it's about the 'software' of the universe, and the mind is very often involved, in one way or another. So the objective reality that might be out there, which I think science needs to find, and will find, is elusive - and for many, not there. That general area is still involved in the mind/matter issues that we now may have the tools to shake off.

So the interactions approach, which to me is the truly promising new avenue of thought, involving the 'hardware' of the universe, not the 'software', often doesn't connect with the quantum information approach.

My view is that if quantum mechanics *doesn't* have some kind of visualisable explanation underneath it, built out of physical analogies, then we may never be able to go beyond a certain point with it. I've always felt that it was worth searching for an intuitive explanation, or a partially intuitive one, in case one exists. Science may only be able to make real progress of a certain kind in the future, if one does exist. But I also found reason to think that one does.

17. An 'exploratory' world?

Talking about the hardware of the universe, another good clue that we have is which of the properties of matter appear when its state changes. It isn't all of them: we know matter in the wave state already has certain properties. But then other properties are only defined in the particle state. So it's worth looking at which properties exist in the wave state, and which ones appear at the point of the switch to particles.

Some dynamic attributes of matter, to do with *motion* such as momentum, only appear in the particle state. But before the wave function collapses and the transformation happens, matter seems already to have static properties like mass and charge. So as you can see, it's specific, but hard to understand. But it's as if something physical is 'going on'.

Because only some of matter's properties come into existence when it goes through this change, the philosopher Jan Westerhoff thinks that the dynamic properties of matter may be mind-dependent, while the static properties are not. In an article called *'Is matter real?'*, He argues that matter may be partly real and partly unreal. This is all very well, but that general kind of approach, which is common among physicists, leaves a gloomy outlook for 21st century physics, as it goes off into mushy mud at the edges of our field.

Carlo Rovelli, on the other hand, has argued that an interaction between two bits of matter is like a measurement. So instead of the observer having any special role, matter is constantly making measurements on itself, whenever two bits of matter collide. So when we make a measurement, we bring bits of matter together to do what they often do anyway. Rovelli says he has no explanation for what's going on underneath that, and it doesn't address the issue about creating routing information. But it conveys a picture of matter as somehow 'feeling out' its relationship to other matter.

Neil Turok has described the wave aspect as an *exploratory* aspect of matter. He said *"[...] our classical picture is wrong. The world is not made up of particles and beams of light with a definite existence. Instead, the world works in a much more **exploratory** way. It is aware of all the possibilities at once and trying them out all the time"*.

To me, this way of putting it again gives a picture of matter somehow 'feeling out' its surroundings, and all the surrounding possibilities. And that picture might also fit with what we know from quantum eraser experiments, where switching the routing information on and off switches between waves and particles. Perhaps if this routing information exists, the setup somehow feels it out, and 'knows it exists'.

It's all very well to think in this general kind of way, but one has to be careful with vague ideas. Neil Turok was talking about the wave state, which he said is 'aware of all the possibilities at once'. That's different from interactions, which lead to the particle state, doing the 'feeling out', or it may be different. And we still have to explain the final choice being partly random.

How do interactions fit with this way of seeing it? Well, they don't really. But nowadays we switch between waves and particles without the experimenter looking. And the thing is, *both* measurements and interactions have trouble explaining that. But we know observation isn't doing it, so measurements go, and interactions start to look increasingly important.

In fact, although incomplete, interactions may be the best lead we have. But if we can't complete the interactions idea, then we might have problems, in the short term anyway. If the very existence of the routing information has an effect, and observations do nothing, then if we can't find a mechanism, to some people it starts to look like the universe might be some kind of mind or computer, or some very complicated system. Many physicists quietly believe that, and some talk about it. Interesting or not as an idea, it could leave our attempts to get to an understanding of quantum mechanics in a place where for now, we can't really go anywhere much at all.

This means that even now, when we've found a way to remove observations from the picture, and even without consciousness involved, science may still be in trouble. The hope is that we can find a far, far simpler answer than that one. And the good news is that in the past, the universe has again and again turned out to be surprisingly, unexpectedly simple.

18. A clue seen only with hindsight

So far this section, Part 3, has been a series of mainly unconnected ideas. I've tried to show a little on how my own thinking, and the thinking of others, has progressed, or tried to progress, over time. I've also looked into some recent ideas. But it seemed good to end the section with something more definite, and I began it by saying I'd set out a major clue in the last chapter.

There's a clue I only saw later, after finding an answer. At that point, I felt it had been staring me in the face for many years. As far as I know no-one else has mentioned it, so perhaps we all missed it. More likely, some know about it, but can't do anything with it at present.

It might provide some food for thought, and will hopefully make good sense as a clue, so I'll give it to you before I describe my solution. It's not the route I took, but it's the beginnings of a different route that goes towards the same picture, whether or not you can get there from it.

In special relativity, we know that some of matter's *properties* depend on the viewpoint from which it's seen. Relativistic energy is one such property, and for many, relativistic mass is another. Over a century, we've got used to the fact that faster moving objects have higher energy and higher mass. But we also know that how an object is moving is just a viewpoint. So oddly enough, the way one looks at an object can sometimes affect its properties, that is, its observed properties.

In Rovelli's relational quantum mechanics (RQM), this happens too. Matter's properties also depend on the viewpoint - straightforward viewpoints that is, like reference frames in special relativity. Now with RQM, this hasn't been proved. But with special relativity it has. We know directly from experiments that relativistic energy is just as it's thought to be. It's hard to avoid the point that matter's properties will sometimes vary with the viewpoint, or frame, from which it's seen. So now you twist that around.

Turning that around, it seems that *without* choosing a viewpoint on matter, it simply wouldn't have certain properties. If so, then in the case of certain properties: no viewpoint = no properties. A viewpoint = properties.

But that can be put alongside something else. In quantum mechanics, part of the mystery is that matter goes from no properties to properties. Before the change from waves to particles, matter doesn't have certain properties. But then they become clearly defined. We know of nothing else that can bring matter's properties into existence.

So perhaps in both cases, the same thing is doing it. Perhaps the same thing is bringing matter's properties into existence somehow. Suppose we look at what happens in quantum mechanics, and ask ourselves a question - *'Do we know of anything else that can bring matter's properties into existence?'* The answer would be *'Well.... it seems unlikely to be relevant, but there is one other thing that does that. Fixing a viewpoint on matter does that.'*

Thinking 'out of the box' a bit, this would mean that in quantum mechanics, when the rapid change happens and some of matter's properties become clearly defined, perhaps a viewpoint on it is somehow being chosen.

If so, it's being chosen out of many possible viewpoints. When I saw that, the question arose: how did I miss that? Well, as I said, something one doesn't expect to find is very often not found. Expectation affects things, and it's an unexpected idea. It's also difficult to go 'out of the box' if you don't yet have anywhere to go when you're out there (the phrase is usually about ideas that already exist). Anyway, for those sort of reasons, I didn't connect two points - two very counterintuitive, but very similar points.

On its own, this clue doesn't give anything approaching a complete answer. More concepts are needed. Others may have found it, but then been unable to take it further. If so, that would be like something that happened to me - Chapter 30 is about how for years I had only half an explanation.

But returning to this clue about matter's properties, it jumps out even more if one looks at matter's state of motion, and rephrases it. In special relativity, choosing a reference frame for motion (a viewpoint) fixes certain properties of matter. And in quantum theory, the collapse of the wave function, which we so far don't understand - - fixes certain properties of matter. Splice those two together, the resulting weird idea is: perhaps the collapse of the wave function somehow *chooses a reference frame.*

It does look like some kind of a choice is being made. This is an unexpected idea because we assume that frames have no physical existence - not of that sort. Anyway, however incomplete this may be, one thing it does do relates to a point in Chapter 16. The question was, what kind of choice in physics, involving choosing out of many possibilities, entirely removes from existence all the possibilities except one? Well, it's quite hard to find that, but choosing a reference frame is a choice of that kind.

And as was mentioned in Chapter 17, often the properties that appear at the point of the change from waves to particles are the *dynamic* properties of matter, such as momentum. Loosely speaking, that's the ones that are to do with matter's relative state of motion. That's about frames. Static properties, like rest mass and charge, exist both before and afterwards.

On that subject, this is a short excerpt from an article by the philosopher Jan Westerhoff, who's at Durham University, UK. In my view, what you see here is him finding one part of this clue (or rather, another clue that suggests the same thing), but perhaps missing its importance.

He doesn't seem to take it as a good clue, because he believes that mind, or consciousness, is part of the process of bringing matter into reality. And that kind of approach (though understandable in the circumstances because it really does look as if that's the case), can make people miss the clues. The article was in New Scientist Magazine, September 2012, and is called "Reality - is matter real?"

"But perhaps this is a bit too hasty. Even if we agree with the idea that consciousness is required to break the chain, all that follows is that the dynamic attributes of matter such as position, momentum and spin orientation are mind-dependent. It does not follow that its static attributes, including mass and charge, are dependent on this. The static attributes are there whether we look or not.

Nevertheless, we have to ask ourselves whether redefining matter as "a set of static attributes" preserves enough of its content to allow us to regard matter as real. In a world without minds, there would still be attributes such as mass and charge, but things would not be at any particular location or travel in any particular direction."

I'd say what he's found there is not about which aspects of matter are 'mind-dependent'. What he's found there is another clue which suggests that when matter changes from waves to particles, what's really being selected is the reference frame.

This also shows that the idea that mind or consciousness is involved doesn't work too well. He points out that even if we assume it is, we then *still* need to divide matter's attributes into two groups - those that are to do with its state of motion, and those that are not. And only one of those two groups can be said to be dependent on consciousness, not the other.

That helps to remove the myth about consciousness, and his reasoning was potentially going in a similar direction, in that he was questioning whether this is the case. To me anyway, this point about dividing matter's properties into dynamic and static ones makes all of this, and quantum mechanics in general, look like it has underneath it a purely physical explanation.

Part 4. Interactions v. measurements
19. Just the ones we know about

For seventy years there was no major progress in quantum mechanics. There was good and significant minor progress everywhere, in many areas - mainly mathematical and experimental areas. We boldly hacked away at the edges of the mystery.

But conceptual physics had to wait, and the conceptual department, where I live, was stuck for a long time. The measurement aspect was hard to avoid, partly because the central theory had been built on it. There were attempts to get around it, but nothing short of a new concept was likely to do it.

But with hindsight, not all that many people knew there was a need for one. The measurement aspect of it seemed to take it all outside the boundaries of physics. And it looked so convincing, and so beyond our scope, that not many searched for a concept to replace measurement. As a result, we didn't find one for seventy years - not until the 1990s.

The two theories that eventually suggested a way to do that, decoherence and RQM, both had roots that went back decades. But they both found their feet and came into their own during the '90s, and that's when they started to change the landscape. This went with a new period of intense debate on how to interpret quantum mechanics, unprecedented since the 1930s, which was partly about new clues that had come in.

But although by the 1990s decoherence was suggesting a major conceptual shift, no-one knew what it meant. And although it looked like it might lead to a solution, the talk about the new idea didn't spread very far, except as an aspect of decoherence, rather than as a crucial, general shift. But the new idea was that interactions are what sets off the sudden change.

Out in the wider universe, the ratio of interactions to measurements is hard to estimate. that's because if there are aliens, we have no idea what they're doing. But it's a safe bet to say that the number of interactions per second is trillions of times more than the number of measurements. And the set of all interactions includes measurements, which is a tiny subset, because to make a measurement, you have to cause an interaction.

And the point is, if interactions cause the collapse of the wave function, then measurements are irrelevant. Measurements is a small group of interactions that we happen to be monitoring. They're just the ones we know about. So perhaps something more general than we imagined is happening across the universe.

If so, the collapse of the wave function wouldn't be an event caused when some blundering creature with a measuring device presses a certain button, accidentally bringing its own reality into existence. Or bringing its own reality into a more concrete existence, out of some background mush. Instead the change to light or matter would be far more common in the universe, and it would happen with or without living creatures.

Now this idea, even before we try to do anything with it, has an enormous ability to resolve existing problems. For instance, before the '90s, it was hard to say how the universe got to its present state, because it was thought that only measurements could help it get there. But we picture the universe's development with a sequence of events. Measurements came rather late in the sequence, when intelligent life had developed (on one planet at least, maybe more). So it's hard to explain how the earlier events in the sequence happened at all.

But the universe had *interactions* going on a long time before living creatures appeared, with their odd measuring devices. So the interactions angle also helps to explain why the universe is as it is now. It helps in many other areas as well, and above all, it offers an objective reality.

So now, "all we have to do" is explain why interactions do that. It seems that we're now only one step away from a sensible solution, but the difficulty of making that step shouldn't be underestimated.

Some interpretations have come near to making interactions the cause, and even Copenhagen leaves the possibility open. But no-one has an explanation with direct cause and effect. Among the new ideas from the '80s onwards, one group are called objective collapse theories. They're good, in my view, because the collapse event is real, and nothing to do with the observer. But then they're not so good, because it happens spontaneously, randomly, and enormously rarely. The general idea leaves it all unexplained, and although it has interactions setting the change off indirectly (by creating entanglements that make the change more likely to happen), it doesn't have interactions as the cause.

Anyway, from the mid '90s onwards, a new view of quantum mechanics was ready to come into its own, if only we knew what it meant. But we didn't, so it never found its way into a lot of places. And the idea that measurements cause the change still was, and is, the standard view.

But that view contains a major problem. Without an objective reality, science isn't dependable, and can't do what it does. We need a reality that exists all the time, and without strings attached - the underlying principles of science, which are based on experimental evidence, demand it. Then we can find out about this reality as we go, which exists independently of us, and is therefore known (although we may affect it a little) to be consistent and reliable in its behaviour. But instead, it seemed that the modern scientific method, which had only been around for 450 years, had already discovered that its own underlying principles don't work.

20. Why didn't we get to the idea of interactions sooner?

But suppose the change that happens to light or matter had been a simple, straightforward, everyday one. Let's imagine, for the sake of argument, that the sudden change when a measurement is made was only an alteration to some basic parameter, like light or matter's energy. If it had been that way, we'd never have decided that our *consciousness* was doing it. We'd never have thought about measurements. We'd have gone straight to interactions, and in about seven minutes, not seven decades. So why did we think about measurements, or observations, at all?

Because matter becomes more clearly defined out of being undefined. That was why, or a large part of why. The change was unexplained, but the main point was, *it seemed to bring the world into existence*, or into a more clearly defined existence. That looked like something exceptional, and suggested we were causing it by observing things.

But if you think about it, that doesn't fit the facts. If we were bringing our reality into existence by observing it, wouldn't it just mysteriously appear out of a strange and unscientific fog of non-existence? Surely it wouldn't need a mathematical set of precise probabilities for different locations? To me this precision, varying with the landscape, suggest a real physical effect.

Anyway, the point I'm making is that our response to the puzzle contained an assumption about what was happening, and our expectation affected it. But in the past, we've often looked at the evidence with less than an open mind. In the Great Debate in 1920, which was about whether the little clouds they could see were in our galaxy, or separate galaxies beyond it, some argued that it was an absurd idea that a nova in Andromeda could briefly outshine the whole galaxy. It was used to argue against the theory that our galaxy was one of many. But we now know that it is, and that supernovas do that.

It's arguable, strictly speaking, that our response to quantum mechanics was premature. We should have waited until we knew more, and simply treated it as we'd treat any effect. It's easy to say that, but we're not robots - we naturally try to understand what's going on. Anyway, nowadays we do know more, because of decoherence, which has allowed us to crack into new areas of the puzzle. We came to those early conclusions naturally, and now, just as naturally, we're beginning to see it differently.

21. The photoelectric effect

But can interactions really replace measurements? One of the best reasons to think they can is in a simple clue on page 86, at the beginning of Chapter 49, which is about how quantum entanglements get started. But here, it's worth saying something very general about the particle state. It supports this picture of things quite strongly: the particle state has a habit of appearing and being observed immediately after an interaction.

In the one-photon-at-a-time double slit experiment, the pattern that builds up on the detector screen is from a series of interactions - with the screen. But because it's a *detector* screen, they might also be measurements, well, they are. We don't know which of these two aspects is the important one. But the fact is, each time a particle shows its arrival at the screen, it does so because of an interaction.

Another example is the photoelectric effect. Einstein found in 1905 that he could show what happens when light knocks electrons out of a metal, using a mathematical setup similar to the one that Max Planck had discovered a few years earlier. He found the only way to describe it that worked, and fitted observations, was to describe the light as discrete blobs of energy - particles - even though at the time people thought light was waves.

But what no-one would have given much thought to at the time though, was that each bit of light that Einstein's new method described as a particle, had always just finished bumping into an electron. It wouldn't have occurred to people that the light's particle nature was being picked out from it, by the interaction with matter that happens in the photoelectric effect. This was before quantum mechanics really took off. It was well before the idea that measurements are a cause appeared in the 1920s, and *way* before the idea that interactions are a cause appeared in the 1990s.

So people will just have thought 'hmm, in that situation, light behaves as a load of particles'. But with hindsight, the photoelectric effect is rather like the detector screen in some versions of the double slit experiment. In both experiments, light travels like a wave until it hits something made of matter, and then immediately starts behaving as particles. With the detector screen, either the measurement or the interaction picks out the particle nature. We don't know which it is yet. But with the photoelectric effect, perhaps we can find out more. If so, one of these experiments might tell us what's happening in both.

In the photoelectric effect, it's not a measurement and also an interaction, it's just an interaction. A light wave hits a bit of metal, and an electron jumps out, behaving as if the light that hit it was particles. It would be very odd to say that a later measurement made on the electron, some time after it has jumped out, acts retroactively in time to make a measurement on the light, turning it into particles (Einstein's discovery showed particle behaviour), so that it can hit the electron in the first place, making it jump out.

This involves a time loop, or a 'stitch in time', as it might be called. It's not necessarily impossible, but Occam's razor, though only a guiding principle, *so* removes it. It's far, far simpler just to say that the initial interaction, when the light wave hits the metal, brings out the light's particle nature, just as the interaction of a light wave hitting a detector screen does.

We could even make the electron that jumps out then set off another event, which in turn sets off a series of further events, in a kind of chain reaction. The measurement we finally make on the last event in the series is then several steps away, but it still tells us if we're getting the photoelectric effect. But the idea that this last measurement *causes* what set all these events in motion gets increasingly far-fetched, with each step away.

Standing back, and looking at the overview of the quantum mechanics puzzle as the 21st century began, to some it looked much as it always had. But this new idea about interactions turns it into a different puzzle, and could make some areas of the old puzzle irrelevant. And yet even by the second decade of this century, the amount of brain time that had been given to the new puzzle was millions of times less. But the fact is, although a lot had stayed the same, one vital area of that unscalable edifice had quietly shifted, and a possible foothold had quietly appeared.

Part 5. Finding new ideas

22. Lateral jumps

It's possible that there's some undiscovered landscape underneath quantum mechanics. Some unseen scenario may lie just behind the clues we have. But if so, then because the mathematics is comparatively simple, it's either going to be something we can't understand, or something we're not expecting.

Otherwise, it would have been found already. Of these two alternatives, one of them is more boring than the other: the idea that we can't understand it, and might have to go on developing our concepts for some time first. But the interesting possibility is that there's some setup that's understandable to us, and that we're capable of finding using our present vocabulary. But if so, it's going to be a lateral jump - something we're not expecting.

But the question of whether it's possible for us to find the concepts, and so solve a puzzle of this kind, is dependent on another question. It's whether an analogy exists at all that we can relate to about it. As I said, if there isn't one, then there isn't a puzzle either, as there'd be no solution. But the universe is full of similarities between things, and physics absolutely depends on that. So hopefully the solution we need will be out there, and it may be a question of whether we can think far enough out of the box to find it.

I'll show a bit about puzzle solving, but I can't say that I used this approach directly on quantum mechanics. I used it on a long circuitous route that got me there slowly, and then saw some of this with hindsight.

A lateral jump is an unexpected sideways move in a train of thought. Edward de Bono identified them, which was in itself a bit of a lateral jump. It helps to know if you're looking for one - for years I used to peel kiwi fruit, which taste good, but it took a long time. Then a friend told me a better way, you cut it in half, and scoop it out with a teaspoon. If I'd known there was a lateral jump to be found, I might have found it. Knowing there's a solution helps a lot. In physics we never know if there's a solution that we're capable of finding.

There's often a false assumption holding you back, that you didn't even know you were making. With the kiwi fruit, it was that you have to get the peel off the fruit, rather than the fruit out of the peel. With hindsight I realised that it used to take ages. But I just assumed that was the only way.

Progress in physics has often involved dropping false assumptions, which we later find out were holding us back. Einstein dropped the assumption that space and time always look the same from all viewpoints - before that they just seemed like the kind of things that do, so people assumed it.

Here are two more lateral jumps and the false assumptions that had to be dropped in order to find them. Before the invention of the rudder (in ancient China), boats were steered by a man at the back with a paddle. It took a very long time before someone thought of fixing the paddle to the boat, with some sort of hinge. Perhaps the assumption that had to go was that the boat and the paddle were two separate objects.

And when basketball was invented in 19th century America, a peach box was used as the basket, and someone had to climb up to get the ball down each time. The ball was later retrieved with a pole, but it took fifteen years before someone realised it could be allowed to fall straight through and back to the players (this may have involved a literal breakthrough as well). In that case, the false assumption was simply that the ball had to stay in the basket, but what allowed the new idea to work was that everyone could see whether the ball went through or not.

There was a lateral jump I didn't make with a tin opener a few years ago, and a false assumption caused it. I found it in a draw in someone's kitchen. Trying to use it seemed straightforward, it had two arms that close to grip, a wheel, a cutting edge - it seemed just like a tin opener. But however I positioned it, it wouldn't go on properly, and it wouldn't open the tin. The assumption that I kept making, and never questioned for a moment, was that the aim is to cut around the inside of the rim on the top of the tin. But that kind of tin opener, I found out later, cuts at right angles to that, just below the top, around the *outer edge* of the tin.

Some lateral puzzles have the kind of answer that seems obvious when you know it, but is far from obvious beforehand. Wheeler believed foundational physics will turn out to be that kind of thing - he said several times that when the 'utterly simple idea' is found, we'll all be saying to each other 'how could it have been otherwise?'.

Some puzzles look like there can't be any solution at all. People sometimes don't even look for one for that reason, and quantum mechanics may be like that. In one of Edward de Bono's books, there's one where three table knives have to be balanced somehow on three vertical wine bottles placed in an equilateral triangle. The triangle's sides (the distances between the bottles) are longer than one of the knives, so it looks impossible to balance the knives on top of them. The solution involves interweaving the three knife blades, so each has another underneath it. Each points towards the centre, but slightly to one side, so interweaving them makes a small inner triangle. My nephews found this fascinating when I showed it to them, partly because they had not believed there was a solution at all.

This is like the invention of the bicycle. Before it was invented, the main thing holding people back from inventing it was the fact that it seems impossible to balance on it. So two separate leaps were needed, one of them to realise it was possible, another to create the actual machine. Two lateral jumps can be much harder to find than one, because if each needs the other, there's a chicken-and-egg problem as well. An intermediate stage helps with getting there, and with the bicycle it seems to have been a 'running machine', which you sit on and scoot along with your feet. That's easier to invent, because it's more possible to realise that it can be invented.

There's also the kind of lateral jump that involves putting something familiar to a different use from its normal one. It's not only James Bond who does this (like using a car tyre to breathe air from underwater), it happens a lot in physics, for instance in the work of Louis de Broglie. Mathematics or physics is taken from one area, and used elsewhere.

One version of that last kind of lateral jump is where one takes an idea that seems to have been disproved, and alters it until it works. When an idea gets ruled out, people sometimes throw out a lot of surrounding ideas, which are actually perfectly viable. When the ether was disproved, all that had been shown not to exist was one particular kind of transmitting medium for light, behaving like matter. But many people then chose to throw out all kinds of possible transmitting mediums, even very different ones.

This is like an eclipse - well, I've always called it that. An idea that is known to be false happens to be near to an idea that's fine, and the false idea blots out the good one. The good idea is then confused with the bad one. So initially, we need to assess each idea on its own, and separate it from context.

In quantum mechanics, not everyone thinks the puzzle can be solved with lateral thinking, but it might be. Just as a footnote, it's hard to define what a puzzle is, but the image of a partly submerged object helps. Some of it is below the waterline, and you have to guess what the rest of the object looks like. It's not all puzzles - some are purely visual, perhaps geometrical, like the ones with metal rings that have to be slid around. But many puzzles involve some structure, system or story that can only be partly seen, and you have to guess the hidden part from the visible clues. Sometimes it makes no sense until you have the answer, and then it suddenly does.

23. Puzzle solving

Those are a few lateral jumps, but I've not shown the process of looking for one, or narrowing things down via the clues. A good lateral jump is an aimed one - you don't just jump off in any direction at all. The thing is to look for the right kind of jump, and that means probing away at the clues first, trying to work out what they're telling you.

Sometimes a puzzle is solved using a 'side channel', which is an indirect route to the answer. A good computer encryption system was cracked recently, by analysing the audible sounds a computer made while decoding an encrypted message. No-one was going to crack the code via the 'front door' - it was too hard to solve. But a side channel provided some vital extra information. This is lateral thinking again, and it's about using lateral thinking to look for clues in unexpected places.

I'll give two examples of word puzzles I made up while hitchhiking around Europe in my early twenties. You wait for ages for a car to stop and give you a lift, but you can't really be reading a book, so I sometimes used to play with mental mathematical puzzles and word puzzles. The point about these two puzzles is that one of them has no clues to go on, while the other has good clues, making a kind of trail towards the answer. So although at first glance one seems far easier to solve than the other, it's actually more the other way round.

The first puzzle is *"find a pair of three letter English words that rhyme, making a full rhyme. Each of these words is the other spelt backwards"*.

This might seem easy, but the problem is, it gives no clues. The relationship between pronunciation and spelling in English is very inconsistent - there are many words that are just spelt in their own crazy way. So looking for three letter words that have endings that rhyme when you spell them backwards is difficult anyway, and one just has to go through a lot of them, trying things out.

The second puzzle looks harder, but it's actually easier, if you know some French. It's *"find a five letter English word that when spelt backwards gives the same word in French, from the same root, but in the plural"*.

(Anyone who wants to try these should do so now, and then read on.)

Now with the second puzzle, if you assume that standard linguistic patterns apply, and some very often do, particularly with words from the same root, then you have quite a few clues. You start by saying that the French word, being a plural, probably ends with an 's', and if so, then the English word starts with an 's'.

Then you say (and all this is much easier with hindsight), that among words from the same root in English and French, where the English word starts with an 's' you often find the French word starts with an 'é', with an acute accent, like écureuil for squirrel, and so on. Assuming this rule applies, the other 'e' can be filled in as well.

So we now have the English word looking like this:

s _ _ _ e

and the French word looking like this:

é _ _ _ s .

Then you say, well they're from the same root, so the two words are similar when read forwards. And yet they're also similar when you read one of them backwards, because the spelling is then the same. So it's going to be a bit of a palindrome, and it seems likely that letters 2 and 4 are the same, in each word. What that means is that four of the six blank spaces above are all likely to be the same letter. There are only a few letters that could fill all four of those spaces (there's only one really), and it's easy enough to try a few.

From there you just guess at it. But the clues, as you can see, have narrowed it down enormously. In fact they've led right to the answer. Two people in the pub in the village in Surrey where I live got the second puzzle, my brother did too, also a friend who's a crossword editor - though it took them quite a bit of time.

I'm not saying that all of the above clues are used consciously, but they help nonetheless. Anyway, in the pub, both Graham and Katie in the end arrived at 'state' and 'états'. But not even Graham, who's a very good linguist, got the first puzzle (we've only ever won the pub quiz with him on the team), the answer to which is 'war' and 'raw'. So the point I'm making is simply that the clues to a puzzle, if there are any, are very important.

24. Economy in assumptions

About a mile down the road from that pub, and seven hundred years earlier, lived William of Occam, in what's now the little village of Ockham in Surrey. I don't know if he went to that pub, or if it existed then - the present version of it was founded two hundred years after his time, in 1540, and the monthly quiz there seems to have started more recently.

William is seen as having invented the well-known principle, widely used in science and philosophy, called Occam's razor. It has had many versions since William's version in the 14th century, and some before. It was later adapted by Newton and Mach, among others.

In our present use of it, Occam's razor boils down to 'In choosing between explanations a particular phenomenon, a simpler explanation is better than a more complicated one, and more likely to be right'. It points out the need to minimise the number of assumptions used in any hypothesis or explanation. Assumptions can be wrong, so the fewer an idea has, the stronger it is, and the less vulnerable it is. Occam's razor often helps to trim away unnecessary assumptions, because if one's choosing between explanations for the same phenomenon, all other things being equal, the one with fewer assumptions is preferable.

In fact, the idea turns up all over the place, and goes back at least all the way to Aristotle. It's only a guiding principle, and not a final arbiter, but the many different versions contained a vital, useful idea. Most of them are concerned with explaining a single phenomenon, but there's an implied link between explanatory power and simplicity. And with a theory that explains more than one thing, if the explanation works, it's arguable that implied in Occam's razor is the following very approximate relationship:

$e \approx p/a$

where e is explanatory ability, p is the number of phenomena explained, and a is the number of assumptions used. The number of phenomena over the number of assumptions gives a rough estimate for various things beginning with 'e', such as explanatory ability, economy, and if it works in other ways, the effectiveness of an explanation. More assumptions, and e goes down. More phenomena, and it goes up. So the object of the game is to explain as many phenomena as possible with the fewest assumptions.

The Irish physicist John Bell used to describe ideas as 'cheap' or 'expensive' in assumptions, because we need to use as few as possible. In those terms, the cost of an idea can be estimated roughly, though it's ultimately a matter of taste. As Matt Leifer once said, 'simplicity is in the eye of the beholder'. The things that are a matter of taste include how to count phenomena, and how to count assumptions, so you can't prove anything with this.

Nevertheless, the explanation set out later, to me anyway, does well in terms of Occam's razor. And it does that whether or not one includes the extra part I've suggested, so whether one takes the things to be explained together or separately. As it happens, the theory was found a mile down the road from Ockham village, when I lived in another village on the other side of Ockham from my present one, though I didn't know about William until later.

Ockham is now just five or six houses that can be seen from the road. There isn't even a pub. I was going to meet another physicist from Surrey for lunch, and he suggested for fun that we meet in Ockham, and asked me what it was like. But it's so minimal, and trimmed down, that there was nowhere to meet there, and we went to another village nearby.

25. Narrowing it down

Some of these points will seem obvious, but in puzzle solving, it helps to ask oneself what one's looking for. Is there anything that can be guessed about what it would be like? It's sometimes possible to narrow things down in this kind of way. With quantum mechanics, the solution would probably include a picture of something: perhaps some structure or mechanism, unexpected and lateral, sitting there underneath the mathematics, underneath the data, underneath all the observations and measurements.

What would it *do*? it would be something that naturally generates the weird but consistent set of rules that apply at that scale. That's the puzzle. Weaker interpretations merely *allow* this set of rules to apply, if one makes various assumptions. The more *generating* of these rules an interpretation does, the stronger it is, and the more of an explanation it is.

Concepts in physics are made of analogies, so what we're looking for is an analogy. In fact, our mathematics is full of analogies as well, but this is about concepts. The exciting point is that being an analogy, if it exists, *it would use some idea that we know about already.* That's a handle on it. One can even look though a list of known ideas that might be appropriate. This might seem almost like cheating, but it's not: there are no rules on how this should be solved. In the '90s I found a way to run a series of questions on what kind of answer it was likely to be, and what came out more than once was (rather unexpectedly) that it would be a clear cut, down to earth, physical, non-woo-woo explanation.

Anyway, that kind of 'digging around the edges' by asking general questions sometimes helps. And the need to look for an analogy is supported by strong reason to think no mathematical breakthrough would provide a solution. In 2011, a near-proof was published that seemed to remove the possibility that quantum theory itself is incomplete. It showed that any future extension of the theory can't increase the accuracy of predictions. So although we may be missing most of the picture, we have the mathematics.

The interpretation that follows is very new, and to me it fits the puzzle well enough to provide evidence for the background theory from which it came. The explanation only works with the background theory underneath it, so it supports the wider picture as well. But the background picture had become too detailed. At the point when I decided to split the book into two books, it had got to 450 pages. Fortunately, the main essence of the interpretation for quantum mechanics can be understood easily in simple form, which is why it made sense to do that.

It's true that an analogy, however good, can't be direct evidence for anything in physics. Strictly speaking, if we decided that an analogy can provide direct evidence for anything, we'd risk allowing intuition to play a part. Because of that, in some ways the very close fit to the puzzle it provides is only indirect evidence. The background theory has the other elements: the mathematics, and predictions for experimental results, which any theory needs.

But quantum mechanics is an exceptional area, in the fact that we have very few analogies *at all* that come near to fitting the clues. So a new picture can still make all the difference. Two points strengthen this later on: Chapter 28 shows how special relativity itself, rather than bringing in new mathematics, was partly a new interpretation for existing mathematics. And there's also Chapter 42, on the vital role of conceptual physics. Anyway, I hope that other physicists who work on quantum foundations will like this view of it.

There are people on the internet who truly help physics to make progress, and one of them is Matt Leifer, at the Perimeter Institute in Canada. He has a website dedicated partly to examining different interpretations for quantum mechanics. His own view is - he's not totally keen on any one interpretation, and tends to think we haven't found it yet. He takes his readers through the the different approaches, keeping an open mind, but expressing his opinion at the same time. I'd say he spreads a lot of understanding, with good clarity of thought in his writing, looking at both the mathematical and conceptual sides, and translating between them. I've seen Ben Dribus taking on a similar role in America.

Whether or not my own view turns out to be relevant, the more we put our heads together the quicker we'll solve these puzzles. A person like that is like a telephone exchange (to use an old metaphor), linking many physicists up, and translating between ideas and different ways of seeing things.

Anyway, Ben, who I've had positive and helpful exchanges with, and Matt, who I've talked to briefly, are very much among the people I've been keen to communicate with about the interpretation that follows. Or to put it another way, I've been dying to show them this.

There's a little more of the background to fill in first, about what we need to interpret, and the landscape surrounding it. Then I'll set out a visual picture, and hope that it works well as an explanation. If it's going to do that, is has to make good sense to people. If it only makes sense to me, then it's not a good explanation.

Part 6. Towards a solution
26. A glimpse of something underneath the water

Quantum mechanics and special relativity have always been like two islands, which seem entirely separate. Until the late 20th century, we knew of nothing that linked them, or joined them up - not in a conceptual way. But they're among our best theories, and they're both well tested. If they're ultimately part of the same picture, they must join up somewhere.

As I mentioned, Einstein and John Wheeler, among the most influential 20th century physicists in America, both said a conceptual basis for physics would be found in the future. The quotes from them about it in the introduction are just two out of many. Wheeler said it very often. Einstein talked about 'the principle of the universe', and said he believed that it would be beautiful and simple. He also mentioned an area where he thought we might look for such a 'conceptual basis for physics', but he was writing decades before relational quantum mechanics. If there was ever an area to look for this picture, it's the place of overlap: RQM.

RQM is *the only known conceptual link between quantum mechanics and special relativity*. It's not just Carlo Rovelli's version, there have been several versions of the basic idea. The basic idea, found by Simon Kochen in 1979, is that quantum mechanics can be reinterpreted as a relational theory.

So if Einstein and Wheeler were right, this conceptual overlap between two of our main theories may be very important. It means RQM might give a rare glimpse of the underlying picture. And perhaps, joining up these two islands, there's something that runs from one to the other. Perhaps it's just possible to make something out down there, underneath the surface of the ocean. But what we see, when we peer down there, is daunting and weird. In RQM, even matter's properties vary between viewpoints.

But then, *we know* both special relativity and quantum mechanics do weird things with matter's properties - and they do it separately. We've lived with relativistic properties, that vary with the viewpoint, for a century, and we've got rather used to them. And in quantum mechanics, we know that matter's properties can suddenly come into well-defined existence. This makes RQM, with its relative properties of matter, easier to accept, and more likely to be right.

It's also worth pointing out that however weird and complicated something seems, complicated effects can have simple causes. So if we can only see the effects, and are trying to guess the cause, it still might be a simple one.

I only read about RQM in 2014, as I'd been working on a very different area. When I did read it, I felt that Rovelli, in bringing these two theories together, had effectively added *structure* to what we see. And there's reason to think that the more structure any phenomenon has, the more likely it is to have an explanation. In the 20th century, when particle physics was being discovered, when a new layer at a smaller scale appeared, the first sign of it would often be a hint of a bit of structure. The bedrock layers of reality are similar. To put it in a basic way, the more complicated something is, and the more separate features it has, the more likely it is that further explanation can be found.

By contrast, spacetime as in special relativity seems very smooth and simple. It looks featureless, and lacking in structure. So partly because of that, many physicists think it's just the way the world is. Perhaps it doesn't need further explanation, and is at the bottom of the well, and as deep as it goes.

But in bringing quantum theory and special relativity together, and showing a way in which they can be taken as similar, RQM added structure. There are now more features, so when we peer down through the water between the two islands, we get a glimpse of some structure underneath the ocean. So it starts to look more likely that there's some sort of explanation.

27. Light at the end of the tunnel?

In special relativity, situations with moving objects are decribed in terms of, and seen through the eyes of, *observers*. But in standard quantum mechanics the same word, an observer, has a very different meaning.

Observers in special relativity are rather straightforward. There's no need for conscious observers. Robots would do. They'd still observe just what Einstein predicted, and they often have, faithfully telling us what they've measured. In the case of special relativity, 'observers' is a convenient way of expressing it. It's really about reference frames, which are particular viewpoints on the universe, to do with how objects are taken to be moving.

But in quantum mechanics, the word 'observer' means something very much more complicated and speculative. It often means a conscious observer, and goes with the possibility that the observer's mind is connected to the world around them in some unknown way. So this is one of the many differences between the two theories. But RQM creates a bridge between them.

RQM brings quantum mechanics closer to special relativity, in the sense that different observers will see unexpectedly different things. And although this is counterintuitive, it's nevertheless a more familiar arena, and one in which we know our way around. And taking quantum mechanics into that arena, this difference between viewpoints is merely about physical viewpoints, like frames in special relativity. No conscious observers are involved. This means that RQM interprets quantum mechanics as a frame-dependent theory, like special relativity. There are no absolute states, only relative ones, such as the state of one system as seen from another.

So just as *velocities* are relative in special relativity, and have no existence at all outside the way they're observed, *states* of matter are relative in RQM, and have no other existence. This even applies to matter's properties, which can be different, if the matter is seen from a different viewpoint.

This is very counterintuitive, and yet it's oddly familiar. Relativistic properties do the same thing. So we recognise something at work here that we already know the flavour of to some extent, and it seems that the universe is up to its old tricks again. Anyway, drawing on a realisation that Simon Kochen had in the late '70s, which was a brilliant discovery way ahead of its time, Rovelli showed that if we take one single weird idea onboard, there's a favourable trade-off. Some other problems then go away, including the worst one - the role of the observer.

So what is the weird idea, exactly? We normally think of there being a world out there, and then you also have the ways in which it can be observed. The two are normally separate, but the weird idea is that there's less difference between them than we thought. So the observables are, for most purposes, all there is. Many people jump from that idea straight back into the idea that reality depends on the observer, but you don't have to jump anywhere quite so quickly.

Suppose instead we move towards special relativity, and away from standard quantum mechanics, and get to that arena where viewpoints and frames are important, but not conscious observers. And suppose we draw a distinction between two things, and say the observer is totally irrelevant, but the angle or viewpoint that matter is seen from isn't.

We're now looking at a new approach, but it's no weirder than some other ideas people take onboard. If there was another reason for this link between the viewpoint on matter, and what the viewed object is actually like, there might be a way forward. Why should matter's properties depend on how it's seen? If there was a reason for that, we might find the observer scrams right out of the picture. That would be good, because the picture was there before observers walked into it, and should be explainable without them.

And there you have what might be a bit of light at the end of the tunnel. The new target is another reason for that setup. For all we know, there might be one. And after all, although this approach is certainly quite weird, it's nothing like as weird as having laboratory men bringing the world into existence with their heads. That's in another league entirely.

So RQM may provide the beginnings of a way out of all these problems. And it's a lot more than just helpful that the weirdness of RQM is reminiscent of other things we know about, from special relativity. The point is, the fact that we recognise something vaguely familiar, even though we don't understand it yet, makes it far more likely to be right.

28. We need an idea

So perhaps this approach provides a starting point for a conceptual idea that explains quantum mechanics. There have certainly been calls for a truly new idea, and Matt Leifer, for instance, says:

'In my opinion, the most likely way that the debate on interpretations can be closed is if one interpretation makes itself indispensable for understanding quantum theory. This could be because it leads to new physics, but it could just lead to a far better way of explaining the phenomena of quantum theory to both students and the general public.'

This, he says, would be like what Einstein provided with special relativity, and shows that Einstein's role was an interpretative one. The mathematics, as he points out, existed already: *'[...] the main advantage of Einstein's approach is that it leads directly to the main phenomena of the theory without having to posit the Lorentz transformations to begin with.'*

So initially, it rather bypassed the mathematics. In making this point, he links to a page where Rovelli also draws a comparison between RQM and special relativity. So a good interpretation should *directly* produce the main physical effects of the theory - and it might bypass the mathematics in doing that. With special relativity, Einstein found a set of postulates that on their own were enough to produce the physical effects. An explanation for quantum theory might do the same, but it should eventually lead to a rederivation of the mathematics, coming from a new starting point.

John Wheeler said he hoped that would happen. When asked in an interview what he thought the best hope for progress in quantum theory was, he said *'Finding a new conceptual basis from which quantum theory can be derived'*. A well-known American physicist, Christopher Fuchs, called for something similar (conceptual, not mathematical) in the abstract of a 2002 paper: *'The suspicion is expressed that no end will be in sight until a means is found to reduce quantum theory to two or three statements of crisp physical (rather than abstract, axiomatic) significance.'* So we need an idea.

The idea that follows can't be reduced to two or three statements, I've tried. Even with four, the sentences are too long, and after a while you start to give up on the punctuation. But with five or six, it gets a bit more realistic.

The connection that RQM has with special relativity means that RQM is very important with or without what follows. But in places, Rovelli's view chimes well with what I'll set out now. It also helped very much with getting there.

29. Real and informational

When you alter the picture, and replace measurements with interactions in it, most of the mysteries are still there, they don't just go away. But it's a new avenue to look down, so hopefully they'll go away further down the line.

The main mystery is the nature of the wave function. That's a question we'll have to answer anyway, whether or not we now have a way to deal with the measurement problem. Returning to the different views of that, two general headings are real and informational. It's worth looking at them a little more closely.

We have a mathematical description of something: it's either a description of a real object, or it could be describing knowledge - or what can be known - about the situation. The two angles are called taking the wave function to be 'ontic' (real), or 'epistemic' (involving knowledge or information). These two views have fought it out in a good natured way for many years.

The wave function is often referred to as psi, because in the mathematics it's represented by the Greek letter psi, or ψ. Those who think it's a real object tend to take the 'quantum realist', or psi-ontic view (someone jokingly called them psi-ontologists recently, because it sounds like 'scientologists'). But with quantum realism, as is looked at in Part 10, you have to be careful with the word 'realism', because it has more than one possible meaning.

Most quantum realists think there's no instantaneous or exceptional event when the wave function collapses. Instead there's decoherence, which is this rapid change, made of many small changes, when the system interacts with its environment. The realist view of the wave as a real, physical wave chimes well with decoherence, which looks like a real, physical process.

But those who think it's about information, the psi-epistemic view, tend to see the wave as representing our limited knowledge of a situation. This view sometimes subdivides into two further versions: one with a real world out there, and the other with no world unless someone is looking at it. These are oversimplifications, but they give an idea.

But the true wave may be some mixture of real and informational, and many have a hunch (as the poll I mentioned shows), that the solution will contain both. This idea of some kind of mixture appears a lot. The early analogy I had for it was the map and territory analogy.

And here's another one. On this idea of a mixture, there's a very good quote from the well-known American physicist E T Jaynes, in which he talks about the mathematics of quantum theory as: *"A peculiar mixture describing in part realities of Nature, in part incomplete human information about Nature - all scrambled up by Heisenberg and Bohr into an omelette that nobody has seen how to unscramble"*.

He goes on to say: *"Yet we think that the unscrambling is a prerequisite for any further advance in basic physical theory. For, if we cannot separate the subjective and objective aspects of the formalism, we cannot know what we are talking about; it is just that simple."*

The formalism is the mathematics. Anyway - some think the wave function is all that exists, some think it exists across many worlds, some think it contains a description of probabilistic knowledge about how light and matter behave, or about some underlying reality. Some think it describes real waves, which affect particles. Some think it's a mixture of reality and information, perhaps somehow scrambled up into Jaynes' omelette. I agree with them about the omelette, my own view is like that (but it has other things in it, as omelettes often do). So now I'll show you where I got to after many years of trying, and have a go at unscrambling it.

30. Two questions

To me there are two questions we need to answer on quantum mechanics, if we're to have any grip on it, or any understanding of it. They're about matter when it's in the wave state. It's easier to talk about matter now, though this can also be about light. When matter is in the wave state, there seems to be a range of possibilities about the situation. And then, somehow, one of them gets picked out.

The first question is: what are these possibilities, what are they possibilities *for*? The second is: how does one of them get picked out?

It seems to me that these two questions must have answers. That's for many reasons, some of which I've gone into already. It isn't enough to say 'that's just the way it is' - there must be something specific going on. But although answers must exist, it's not certain how comprehensible to us they'd be. We might have to go on developing concepts for some time first. Still, the clues are simple, and I felt that there absolutely must be reachable background concepts, and that when we have them, however long that takes - then yes, there would be answers we can understand.

In 2003 an answer for the first question jumped straight out of a theory I'd been working on since 1995. The theory involved a new way of defining the structure of space, and was like a background picture. The idea for quantum theory was one I should have seen before that point, but it was a lateral one, and ideas like that are sometimes hard to see. So from 2003 on, I believed I knew the answer to the first question, but not the second.

That seemed very encouraging. But an answer for the second question didn't come. Months went by, and after a while the months turned into years. Then the years turned into eleven years. For a long time I believed I knew what the possibilities actually were, but not how one of them is picked out. I kept on searching, because half a solution isn't enough - it's no good having half an explanation for something. The first half was simple, though it came out of a group of lateral ideas. That part seemed certain to be right, and even more so over time, so it seemed increasingly certain that there'd be a second half out there. But I couldn't find it. After a while it got rather left to one side.

Then one day in 2014 I read something that Carlo Rovelli wrote. That year I'd been reading about RQM for the first time, after coming back to quantum mechanics, but I hadn't found this particular bit of Rovelli's view yet. He said he didn't know why, but that he thought an interaction between two bits of matter causes what he called 'an exchange of relational information'.

When I read that, something clicked into place, because I knew exactly what I thought that relational information was. To me, in the particular context of the picture I'd been building up, Rovelli had come up with a way of seeing it that landed a vital handhold onto what was really going on. So from there, putting an idea of his into the picture I had, I got to an answer for the second question.

Part 7. A possible explanation
31. The first question

So now I'll tell you my solution, and then I'll fill in more of the background. That's partly about the theory that this picture came out of, which connected with Rovelli's way of seeing it in 2014. It felt a bit like when they met in the middle while digging the channel tunnel, and not because he lives in France and I live in Britain. When I met Carlo Rovelli in 2017, to talk about quantum mechanics, I found that rather like with the channel tunnel, we were coming at it from very different directions.

I should mention that there's disagreement among physicists about attempts to visualise quantum mechanics, and what we can say about the wave aspect of light or matter, with statements such as 'the photon passes through both slits'. Some criticise descriptions of that kind as mistakenly using concepts from classical physics. But we don't know. And if there *were*, in fact, some comprehensible setup underneath quantum mechanics, how would we ever find it? If we don't allow ourselves loose, initial attempts to visualise what's happening, how would we ever get to the right picture?

It's like an analogy that Ian Stewart and a collaborator came up with: if every new car had to be better than a Rolls Royce with the first prototype, no new car designs would get off the drawing board. This really means 'allow for the process of development'. I said something a bit similar to Neil Turok about conceptual thinking, and he said he tended to agree - that you start with loose fitting concepts, and then later on you tighten them up. That's how the ideas that follow got started. But also, when communicating a visual picture, it helps to start with a simple version, just to get it across. So I hope you'll be prepared to enter the spirit of picturing these things loosely.

So I'll begin with the first question. What *are* the possibilities in the wave function - what are they possibilities *for*?

According to this view of quantum mechanics, the set of possibilities we see, is really about another hidden set of possibilities we don't see, sitting there right underneath the ones we do see. They're to do with *the only other thing we know of*, apart from the wave function itself, that can be in a physical superposition, with many possibilities that co-exist. It seems to me there's only one other thing that does that.

Well, I'm talking about the axes of the dimensions. We've taken them more literally for a few decades. And if you do take them literally, then they're in a physical superposition, because they're placed at many different angles at once. This creates a set of possibilities, about how the dimensions might be positioned in space. It's about their orientation, that is, the angle at which the axes are orientated.

What we call the dimensions is a set of axes in space, with no fixed positions. Forty years ago we took them less literally than we do now. But whether you take them literally or not, the basic idea is the same. Unless a positioning for the dimensions has been specified, they're not set at any particular angle. Or alternatively, they're set at many angles at once. If it's many, then they're in a physical superposition. Either way, there are possibilities about it.

In my picture of quantum mechanics, until a positioning for the dimensions has been established, they have no fixed positions. Instead, there are many possibilities about how they might be angled. But exactly how they might be angled, and where they are, is important for matter, because in this theory, matter is places where the dimensions are vibrating.

32. The structure of the dimensions

Nowadays the structure of the dimensions is thought to have more to it than just three straight lines at right angles, coming out of a vertex. Those three familiar axes, which are called the flat dimensions, can be angled in any way one chooses.

But particularly since the '80s, because of string theory and other theories, it has been widely thought that there are also other 'compactified' dimensions. They're often seen as curled up into small circles, making very thin cylinders at a small scale. Each cylinder is made of a circular dimension and a flat one, and loosely speaking, it runs along the straight line made by a flat axis.

So in that sort of picture, at a very small scale, space has a kind of grain to it. In this theory, if it were possible to zoom right in, space would look like a lot of parallel cylinders. But how they're angled is often undecided.

A basic unit of matter, in this theory, looks a bit like the closed string in string theory, but it's very different in some ways. It's a small circle of waves in the fabric of a cylinder, travelling around it.

But the dimensions make a single structure, and because the positioning of that whole structure can be undecided, the orientation of the plane of this small circle will also be undecided. It depends on how all the axes are angled, which affects the position of the cylinder, and that in turn affects the circular object that lives on it. So the way the dimensions are positioned is important, because it affects matter's orientation in space.

As a result, it's difficult for matter, because what happens to it depends very much on where the dimensions are. Matter exists on the dimensions, and is a part of them: it consists of ripples in their fabric. So matter has to go where they go. But the trouble is, sometimes it isn't clear where they go.

33. Dimensional quantum mechanics

This theory is called dimensional quantum mechanics, DQM. One key point in it is, some things about matter depend on its *orientation* at the Planck scale, relative to the system (the chunk of matter) from which it's observed.

There are things that depend on matter's orientation in space. Even some of matter's properties do, but that will make more sense later on. For now, the general point is that the relative orientation of the circles of waves can affect what matter looks like when we observe it - what configuration it happens to be in, and so on.

And it also affects where matter might be located. Imagine an electron flying around the nucleus of an atom. Or rather, doing whatever it does there, we don't quite know what it does. We know that instead of having a location, its location will be 'smeared out' into a probability wave. At any given moment you can work out, within a volume of space (the size of the atom), where it might be, and how likely the different places for it are, using the Schrödinger equation.

But you simply can't know exactly where it is. Its undefined location starts to make sense if the electron is really a disturbance on a circular dimension, far smaller than the space in which it might be found. So it can be imagined as a very small wave in the fabric of space itself, running around a cylinder made of space, in one particular place on it.

If that's what the electron really is, then until we know how the dimensions are angled, including the cylinder on which this particular electron lives, we don't know where the electron is. Instead, we just have a set of possibilities about where it might be, and we can work out the odds. The odds depend on the situation, and some locations are more likely than others.

Let's imagine all the places where one of the three ordinary flat dimensions might be. The possible positions for it are straight lines that radiate outwards at different angles, and they come from a vertex, which is a point at which they all join up. This can be at any point in space, but there's reason to think that within an atom it can be taken to be at the centre, at the nucleus.

Now the dimensional cylinders are very thin, and run along the flat axes. So the cylinder on which our electron lives also has a set of possible positions, and they also radiate outwards at different angles. At any given moment the electron is at a specific place on the cylinder, so each possible *angle* for the cylinder implies a possible *location* for the electron.

This would explain the strange set of possibilities that we find. They're about angles, but what we imagine is locations. We already know that the electron behaves in an unexpected way. But it goes by a precise set of rules, and they have their own weird but self-consistent internal logic. We usually take those rules to be just about the electron, but they might not be.

And if we put in the background setup of DQM, and make the electron a part of that picture, it starts to become possible to explain its behaviour. This will hopefully get clearer as we go, but perhaps you can see the beginnings of the picture.

34. The second question

But to explain the electron's behaviour in a more complete way, we're going to need to answer the second question as well, which as I said, gave me a lot of trouble. It did until my incomplete picture connected with something that came from Rovelli. He had got to a part of the picture that I couldn't get to, and in an entirely different way. And he had got there without even believing the wave function necessarily exists at all, even though most physicists think it exists in one way or another.

Several years later, I found out that although putting together an idea of his with ideas of mine led to an interpretation for the wave function, he leans towards a view in which there's no wave, no wave function, and no quantum state. Instead there are places where matter becomes visible for a very short time, but between these points, its paths can't be traced.

Rovelli got to the key idea of interactions via Heisenberg's work, which in the 1930s was a parallel, alternative approach to the Schrödinger wave function, and very different. He may also have been influenced by decoherence, as I was. But it was RQM, his own relational theory, that got him to 'an exchange of relational information'. That particular concept, when put into my picture, made a way forward appear.

It was about what happens when we make a measurement, for instance on the electron, and find out exactly where it is. We know the electron *only has a location at all* once we've made a measurement on it. Before that point, its location takes the form of a set of possible locations. That part I felt I could already explain, including even the superposition itself. But this is about the second of the two questions, which was: how does one of these possibilities get picked out? That question gave me a hard time for many years.

Standing back a little, the electron is in a laboratory (if not we wouldn't know anything about it), and has been waiting to have a measurement made on it. Before we go any further, there's a need to think about the laboratory, and what it contains.

35. The laboratory

The lab is at a different scale from the quantum world, and we know that for some reason, the large-scale world behaves differently. In the first half of the 20th century, people thought there might be two different sets of rules - one set of rules for the large-scale classical world, and another for the small-scale quantum world.

But nowadays we generally think there's just one set of rules, and the world is quantum at all scales. That's partly because we've learned more, and also because rules should be universal. But if there's one set of rules, it seemed to me there must be something emergent going on at a large scale, to explain the transition. I'm one of the people who believe that, and I'd say that if we don't find something emergent to explain it, it's not clear how we're going to explain it. What follows is a description of an emergent setup of a kind that can interpret the scale issue in quantum mechanics.

Decoherence has shown a lot about the transition between these two scales, and the good clues from that area are, as always, the ones that have allowed our view to change and develop.

So back to the electron. In DQM, when someone makes a measurement on the electron, they have to cause an interaction with some light or matter in the laboratory. And the lab is at a much larger scale than the quantum scale, and because of that, it already has an implied positioning for the dimensions. This positioning for the dimensions arises from the relationships between the bits of matter that are sitting there in the lab. These relationships have already been established via many interactions.

The links between the bits of matter in the lab cross-corroborate each other, and the resulting positioning for the dimensions is unambiguous. Something like a framework builds up, which creates the large-scale world as we know it. And (as becomes clear from looking at decoherence), many of these links are entanglements - more of that later. But at a large scale, loosely speaking, there's a pre-established positioning for the axes, coming out of an already existing network of spatial relationships. It's a purely relational thing, and is emergent. But like many things that are emergent, it's very real.

So there's what might be called a 'local framework' in the lab, and when a bit of matter bumps into light or matter that's already part of that structure, it gets its bearings in relation to the lab, and can join onto the framework.

In fact, it's not quite that simple, for several reasons. All matter is in motion at many scales, for one thing. So the matter in the lab shifts around, and bits of matter move in relation to each other. The motion creates a setup that's more complicated. But then, photons come flying through the lab, bouncing off what's there, carrying information about what's where, and keeping the framework updated and maintained via interactions. So it's frequently reset, and at a large scale, there's something that matter can latch onto.

But at a small scale there are bits of matter floating around that haven't yet bumped into any particle from the lab, so they're still in a superposition of states. They haven't yet 'got their bearings', so their properties take the form of a set of alternative versions of the situation.

But what about the electron, what happens when someone finally makes a measurement on it? When the measurement is made, the electron interacts with light or matter from the lab - that's what a measurement does. Colliding with something that has already angled itself to the lab framework gives the electron what Rovelli called 'relational information', which in DQM means, more specifically, orientation information. That interaction provides a spatial orientation, relative to the lab, for the cylinder on which the electron lives. And as I said, each possible angle for the cylinder implies a possible location for the electron. So when the cylinder finds out its orientation, one angle is picked out, and the electron find out its location.

The next chapter has an explanation for the wave-particle duality. But I have to fill in a little more background first. In the background theory that DQM comes out of, you have a lot of parallel cylinders, made of space, at a small scale. Light and matter are disturbances in their fabric. And loosely speaking, matter travels around them, while light travels along them. But a light wave is at a much larger scale, and ripples through many of the cylinders, so this is just about the *direction* in which things travel.

But with matter in DQM, in a frame in which it's moving, it also moves along the cylinder it lives on. So both light and matter, when they move in ordinary three-dimensional space, travel along the length of the cylinders. And when they do, the cylinders are aligned with the direction of motion.

The result of this setup, which is gone into in detail in Book II, is that in DQM, matter can only relate to other matter, in any way, if there's a positioning for the dimensions. Even just the idea of relating matter to other matter tends to involve one, and if you *are* relating matter to other matter, before you know it, you've implied one.

36. The wave-particle duality

The dual nature of light, as waves and particles, has been a mystery since the early 20th century. At that point, we found out that it's somehow both. But in the centuries leading up to there, it looked like it'd be one or the other. So before that, the general view often swung backwards and forwards between waves and particles, as more clues were found, and people wondered which it was. No-one thought it could be both. Newton thought light was particles, but his contemporary Huygens thought it was waves.

But by the early 20th century we knew that light has a dual nature, and we've been struggling to understand it ever since. And because *matter* turns out to have this dual nature as well, it's everything we can measure, and everything we can observe. So to explain the wave-particle duality, one would need to explain a central aspect of the universe.

But to do that, there would be a need to explain why this duality affects both light and matter. They're different in some ways. Matter has mass, and can interact with itself; light doesn't, and doesn't. But they also have similarities, and the wave-particle duality affects both. So to explain it, there'd probably be a need to show that this duality arises at a more fundamental level than where the differences between light and matter appear.

In DQM, what's behind the dual nature of light and matter is the dual nature of the dimensions. To me, the dimensions are the transmitting medium for everything. They're the underlying bedrock from which the world emerges, and everything is made of disturbances in their fabric. So if *they* have a dual nature, it'll be basic enough to make a difference further up, and will be able to affect both light and matter.

At this point it's easier to talk about matter. So matter has two states, waves and particles. The underlying reason for that, in DQM, is that matter is part of the dimensions, and the dimensions themselves have two states: without defined positions, and with defined positions.

When the dimensions don't have defined positions, the flat axes make many straight lines, all set at different angles in space. And as I said, when light and matter travel, they move along the flat axes. So when a particle moves along *one* of the axes, because that axis is reproduced in many places at once, the particle moves along all these different paths at once.

And that's what makes the wave. An emitted photon spreads out into a wave because it takes all these possible paths. The photon becomes a spread-out object, as seen from our vantage point, or a wave comprising many different versions of the photon. To picture the wave, it's like seeing multiple images of the same object, all overlapping. But they're not images, they're real.

By contrast, when the dimensions *do* have defined positions, instead of the action happening along many lines, positioned at many different angles, the action happens along just one line, at one angle. And that's what leads to the particle state.

This point about multiple images is a bit like an argument that broke out in astronomy in the 1970s. Sometimes a large galaxy would be found to have a number of smaller galaxies surrounding it, all very similar, and some even thought they were being born out of it like babies. But later on it turned out that what was being seen was several images of a single more distant galaxy. The light from it was being lensed by the nearer galaxy, and taking different routes, and this put the image seen from Earth in several places at once.

In a very loosely similar way, I think a quantum wave can be broken up into what is like a set of multiple images of the same event. Because one particle travels at many different angles at once, it's as if it's being seen through a sort of 'multiplying lens'. But because each of the particles is real, instead of saying multiple images, it's better to say that we're seeing multiple versions of the same event.

So that's why, for instance, a given particle can be in more than one place at the same time. It's because the dimension on which it exists can be in more than one place at a time. And that idea isn't just a part of DQM theory. What we call a dimension *really is* in more than one place at a time.

This view of the wave state chimes with Feynman's path integral approach to quantum mechanics, with which he provided a glimpse of what's going on. Taking a unique, oddball angle, as he often did, Feynman discovered seventy years ago that very similar mathematics could be arrived at by assuming that a particle travels on all possible paths at once. The roots of this work go back to Dirac and others, and the idea has been important in subsequent physics. But at the time, like many other things, it just seemed weird.

While talking about Richard Feynman again, I should say that he worked with John Wheeler, having quoted and mentioned both of them. The two worked together on absorber theory, and many other things. Feynman once said an idea he had that led to a breakthrough came out of something Wheeler said to him on the phone. Feynman was originally Wheeler's pupil, and they were described as more or less opposites in temperament - Feynman was seen as an oddball eccentric, while Wheeler came across as a reserved gentlemanly academic. But they both had very original ideas, and sparked each other off, regardless of their differences.

37. Quantisation

Quantisation is one of the most basic elements of quantum theory, and what gave it its name. It was also the first glimpse we had of the theory, and came out of Max Planck's efforts at the end of the 19th century to solve a puzzle of the day. The challenge was to explain how energy is radiated by a hot object, in a way that made mathematical sense. He eventually found it was possible if he divided the energy into small, equal chunks, and it was later found that the world is like that in many places.

This simply couldn't be explained by classical physics, in which the waves are continuous, and there's an infinite range of values between one number and another. The initial problem was the 'ultraviolet catastrophe', which was that classical physics breaks down when it tries to describe emission at very short wavelengths. What this really showed was that the classical description was an approximation, and needed replacing with something better. But no-one thought it would take a revolution to sort it out.

Planck worked at the problem with great intensity, trying almost anything he could think of before he solved it. When he did, although he didn't realise it at the time, he opened the door to the next paradigm.

Ironically, Planck had been advised not to go into physics at all - he was told the field was more or less already sewn up, and that there wasn't much left to be discovered. (This kind of view arises quite often, as people sometimes paper over the cracks. But Einstein, and many with the same attitude as him, are always found staring into them.) Planck said that he would study physics anyway, as he just wanted to learn what was already known, and had no aim of contributing anything new. Among the range of things he discovered was quantisation.

And it's still unexplained. Rovelli once said he thought that any interpretation for quantum theory that doesn't explain quantisation is not worth looking at. A light wave's energy can only be divided up only discrete units. The size of a light quantum depends on, or is related to, other things about the wave, and as usual, we have the exact mathematics.

Perhaps you can see how DQM explains quantisation. The chunks are utterly indivisible within a particular wave, and simply can't go any smaller. But they nevertheless can be found smaller elsewhere - if one turns ones attention to a different wave. So there's no *universal* fixed unit of energy, as one might expect. Instead something about the wave creates it.

According to DQM, light's energy comes in discrete chunks because the wave consists of the same photon seen many times. The wave is a single particle, but seen spread out, so it exists in many places. So each wave is based on its own particle. Two separate waves are based on two different particles, and because of the relationship between energy and wavelength, if the particles have the same energy, the two waves will have the same wavelength.

This picture explains why the only way to find smaller packets of energy is to look in a different wave. The other wave will be based on a different particle, with a different energy.

DQM explains the energy levels of orbitals within an atom in a different way. The above only explains the relationship between waves and particles within a quantum wave. But it applies to matter as well as light, and still provides a general explanation for the nature of a quantum wave, involving the wave as many superimposed versions of the same particle.

It answers, among other things, what's really the main question: why a given wave can only be divided into fixed and equal units. If these units are really different versions of the same object, then it's not surprising that they're all fixed and equal.

Part 8. Some other phenomena and concepts
38. The Planck scale

What is a dimension? A lot of things in physics are defined in a rigorous way, but dimensions are not rigorously defined. Still, the main things about them are well understood, and perhaps the fact that we have no exact description means there's more room for them to surprise us.

In the standard way of seeing them, the dimensions are possible directions in space. When they're taken more literally, space has a structure to it, and the possible directions are built into it. When they're taken less literally, they're lines we draw in space to help with our mathematics. In fact, we've moved from the less literal version to the more literal one, and what happened was, the lines we drew in space turned out to exist. This was a surprise, but then we never drew circles, and there seem to be small circles as well.

Anyway, taken literally or not, the dimensions still show something about the nature of space. Either way, they contain something real about the universe. They do have positions, but only in relation to each other. They go by certain rules - the main thing they have to do is to stay at right angles to each other.

The Planck scale is another of Max Planck's discoveries. It's the smallest scale we know about. In the 1920s people found reason to think there might be an extra dimension down there, curled up to a small circle. But the dimension was impossibly small, and until the 1980s, not much could be done about it. Then string theory took off, and curled up dimensions started looking a lot more relevant. It's now widely thought that they exist.

But Planck scale physics hasn't been used In attempts to interpret quantum theory, until now. It's a separate area, they're at very different scales, and it's a lateral jump to bring in one to help explain the other. To give an idea of the size, very roughly, the Earth is 10^{20} times smaller than the visible universe. A large nucleus of an atom is 10^{20} times smaller than the Earth, and that's near the particle scale. The Planck scale is the same factor difference away again, so it's *another* 10^{20} times smaller.

These last two scales are not only different in size, they're also different in accessibility (how gettable at they are). Nowadays we do experiments easily at the quantum scale, and have got very good at manipulating what we find there. But at the Planck scale, there's no chance of doing experiments at all. We only have one tool that can reach down into that strange, hidden world: mathematics. That is, apart from the human imagination.

But despite these differences between the two scales, we have no complete picture of what's going on at either of them. So for all we know, events and behaviour at one scale might be causing events and behaviour at the other. And in fact, many believe this to be true anyway, as that's what happens in string theory. So it's far from being an outlandish idea.

Bertrand Russell's version of Occam's razor added the point that the more an explanation contains concepts that are already on the map, the better it is, and the more credible. He said *'Whenever possible, substitute constructions out of known entities for inferences to unknown entities.'* So if quantum scale physics can be explained using Planck scale physics, that's good. When trying to fill a sizeable hole in a jigsaw, it's good to make use of the loose chunks of jigsaw, that we've already put together.

39. Two choices that were always seen as separate

I've mentioned that in DQM, when light or matter move in three-dimensional space, they travel along the length of the cylinders. The cylinders are aligned with the direction of motion. This means that if you want to relate one bit of matter to another in any way at all, you'll be deciding something about how the dimensions are angled locally.

It's like choosing a reference frame. In DQM, when choosing a frame, one is also automatically choosing a positioning for the dimensions, or limiting the possibilities for that. This suggests a basic link between two arbitrary choices in physics, that have always been taken as separate. They're about choosing a frame, and choosing a positioning for the dimensions. In the past it would have seemed odd to talk about connecting the two choices. Before the '80s, the dimensions were usually not taken literally enough for that - they were often seen as a useful mathematical device.

Reference frames have not been seen as particularly real either, for obvious reasons. They're viewpoints on the universe. But in DQM, when we choose a reference frame, and decide how one bit of matter is moving in relation to another, we imply a relative orientation between them, or a positioning for the dimensions. It turns out that in DQM, the relative orientation between two objects, as well as having bearing on their relative motion, affects their observed properties as well - which are also relative.

Now a frame of reference, as already we know, can be chosen in a university. But we didn't know one can be chosen out in the universe. If it's the former, it might be for a calculation. But according to DQM, choices like that are also being made across the universe, in a far simpler way.

When two bits of matter collide, there's what Rovelli called an 'exchange of relational information'. To me that's an exchange of orientation information. It establishes their relative orientation in space, and that information carries other pieces of information with it. These include the relative motion of the two systems, their properties (for each as seen from the other), an implied positioning for the dimensions, and so on.

And that can include a frame. So unlike the choice of a frame by a physicist, which doesn't affect much except the calculation that he or she is doing, the choice of a frame via an interaction actually affects things. It shows that what we call frames of reference can be more real than we thought, because they can be established locally, as part of the emergent level in the universe.

40. A very direct property

Before anyone makes a measurement, we know matter doesn't have certain properties. When a measurement is made, its properties somehow become clearly defined. This strange fact was what made measurements seem to be bringing the world into a more concrete state of reality.

So how can matter *not have* certain properties? This is rather reminiscent of those odd conundrums (like 'what is the sound of one hand clapping?') that are more or less meaningless if taken literally. But by contrast, this question *is* to be taken literally.

In this picture of what's happening underneath quantum mechanics, some of a particle's properties will depend on the angle it's seen from. As is set out in more detail in Book II, some properties are not frame-dependent, but others depend on the positioning of the dimensions, and so on the viewpoint. In both RQM and DQM it's all about viewpoints. Rovelli's approach fits with my version of things in several ways - at the time it brought clarity that simply wasn't there before.

The RQM picture includes the idea that an object has fixed properties from the viewpoint of another object, if the two systems have already interacted. So before relational information is created, the *lack* of properties of matter that we know in quantum mechanics is itself part of a relative setup - RQM relativises even that.

But before RQM, many of us thought that this definedness or undefinedness of matter was the same from all viewpoints. We assumed that when matter jumps into focus, it does so for anyone watching from anywhere. Well, I had assumed that, though without ever actually thinking about it, as you do. So for me, RQM became part of a new way of seeing things.

This dependence on the positions of the dimensions for matter's properties can be seen very easily in the case of spin orientation. We know that pairs of entangled particles can have the same orientation in space (their spin may be in opposite directions, for instance), even if this common orientation they share is undecided. According to quantum theory, this orientation will only be defined *at all* after a measurement is made. In the DQM picture, when the measurement is made, the interaction that has to happen to make the measurement establishes where things are.

Now according to DQM, there you can see the mechanism at work. This time we don't have to *infer* that matter's properties get fixed by establishing its orientation in space. Because spin orientation is a very direct result of what happens, for once, with this particular effect, we can actually see a property of matter being linked to its orientation in space.

That goes without saying really, the property *is* its orientation. But the point is, we already know that a measurement (or an interaction) can fix matter's spin orientation, and that the property then appears *out of non-existence*. So the fact that this can definitely happen for some reason, strengthens DQM's main premiss - that what's actually being decided at the point of the sudden change is matter's orientation in space.

41. Layers of interpretation

RQM in its present form, or something approaching it, was found in the mid '90s. I only read about it years later, as I was looking at other areas, by which time it had become well-known. Other early '90s versions were similar, and the idea has roots that go back to a 1979 preprint from Simon Kochen, an American mathematician originally from Belgium, who nowadays has been at Princeton University for around 50 years.

He later put the same idea forward again, in a conference paper in 1985. As Kochen put it in his now well-known paper: *"The basic change in the classical framework which we advocate lies in dropping the assumption of the absoluteness of physical properties of interacting systems [...] Thus quantum mechanical properties acquire an interactive or relational character"*.

Kochen's work is enormously important, and he made the first mental leap towards RQM on his own in the 1970s, way, way before anyone else saw it. Amazingly, all the main pages about him online don't mention the discovery at all, but only less important things. Others put Kochen's idea into different forms, and Rovelli's became the widely known version. But there are other versions, from before and after Rovelli's 1994 paper.

There are some similarities with special relativity. Rovelli draws a comparison between them, pointing out that Einstein brought in a relational element to space and time. Rovelli says his own work does the same kind of thing with quantum mechanics. In both cases, the mathematics largely existed already, and what was added was a relational interpretation.

To me - and Rovelli has hinted that he thinks the same - relativising quantum mechanics leads to an 'intermediate level' interpretation. It says something about what's happening, but not why. RQM has always potentially needed a deeper interpretation, explaining the relational aspect, and why a number of things about matter depend on how you look at it. DQM provides a deeper level interpretation, and hopefully shows what's going on underneath both the layers above it. That is, the effects of quantum mechanics, which is the top layer, and RQM, which is the intermediate level.

42. Conceptual physics

Conceptual physics is a vital area of physics. It's what allows us to make real progress. Conceptual physics isn't metaphysics or philosophy, although it has sometimes been mistaken for them. (It's also not the teaching method of the same name mentioned on Wikipedia.) John Wheeler talked about it a lot, for instance in a filmed interview on 'Web of stories', where he said: *'It's not just a matter of nice simple formulas, there's some ideas out there waiting to be discovered.'*

The conceptual side is arguably the very backbone of physics. Some excellent physicists, and not only Einstein and Wheeler, have seen it as central. Victor Weisskopf was present at the birth of quantum mechanics, and worked with some of the main people, including Nils Bohr. He was later a much loved and respected Professor at MIT, who through a distinguished career spoke out for conceptual physics. His well known phrase was 'Search for simplicity', and he used to say his old teacher, Paul Ehrenfest, had influenced him to think in a conceptual way.

Some philosophers also see a need for more than just mathematical theory - Bertrand Russell, for instance, thought that the objective of both science and philosophy was to *understand* reality, not just to make correct predictions on it. Many physicists take the middle ground, and think of the conceptual side as one medium sized part of the whole process. But a small number decided it was irrelevant, after it started doing badly in the 20th century.

But that wasn't its fault. It's just that we weren't ready to interpret some of the mathematics we found in the 20th century. We needed to make progress in other areas before it was possible to get there. And quantum mechanics was so counterintuitive that it was a very legitimate question to ask if it even had an interpretation. We didn't know. But the point is, there was a need to suspend judgement. But instead of keeping an open mind, as the picture had got behind theory, some decided to remove it.

But conceptual thinking has led to almost every major breakthrough that has been made in physics. It's often about making *connections* between things that had previously been seen as separate. In this book, you'll see me trying to splice together pairs of concepts that are not usual dancing partners. And whether or not it works here, in the past it has worked to do that.

Newton made a connection between the force that pulls the apple, and the force that pulls the moon. He realised that the two might be the same force. A calculation then showed they matched ('pretty nearly' as he put it), but the idea came first. Einstein made a connection between free fall and no gravity, which was the first domino of many, and led to a series of realisations.

So the conceptual side is always right there when the key steps are taken - at the cutting edge of our understanding, and showing us where to look for the much needed mathematics. By contrast, searching through mathematics on its own, and trying to find a breakthrough that way, usually doesn't work. It's like trying to do a jigsaw in a darkened room: one can test how the pieces fit together, which is like the mathematics. But there's a need to see the picture as well.

But during the 20th century, partly because we couldn't get the concepts we needed, a minority decided that only the mathematics of a theory counts for anything, along with predictions. And yet it's clear that if at some time in the future we found an underlying picture (as some of our best physicists said we will), our view of conceptual physics would undoubtedly change. So although some at present think little of it, that view may go anyway.

And the majority know there's more to physics than mathematics. Very often the predictions, and other important elements of a theory, arise partly from the picture. The quote from Christopher Fuchs (on page 54) uses the word *'physical'* to mean conceptual, and *'abstract, axiomatic'* for mathematical. This shows the deep link physics has with concepts.

I don't like to speculate about DQM being confirmed in one way or another, or accepted for whatever reasons, but one point can be made. If the DQM picture is right, it would shift things in physics towards the conceptual side, making that side more important. It would mean that conceptual physics was important all along, and as a result this picture would then be more relevant, even without the accompanying mathematics (which is in Book II, with the more detailed version).

So it would be indefensible to try to take credit via a more complete version of the mathematics, as some have tried to do with other theories in the past. And if DQM is right, trying to do that would be ironic, given that ninety years of attempts to solve quantum mechanics via the mathematics failed, and the problem was finally solved via conceptual thinking.

Anyway, with or without that kind of shift, conceptual thinking was always a key part of the discipline. It not only led to the main breakthroughs, it's also central to the methods and goals of physics in a more general way. One point worth looking at is the question of what our physics would have in common with that of another civilisation, if such a thing existed.

There may or may not be other civilisations, built by intelligent creatures on other planets. But suppose there are such things, and suppose we met some of them over the next few centuries. It's interesting to ask what our physics would have in common with theirs. We can't say much about our advanced physics, because we don't yet know how fundamental our present ideas are. They need the test of time. But going back to basic, well-understood physics, that has already gone through the test of time, it's quite easy to establish a few things. And what comes out is that the mathematics might be different, but the concepts will be the same.

Take heat, for instance. These creatures might have mathematics to describe heat that's very different from ours. There's a lot of equivalence to be found in mathematical physics, which means that there are plenty of different ways to describe the same thing.

But the creatures are still going to agree with us, in a basic sort of way, that when an object gets hot, the bits of matter in it are moving around more. It's a fundamental part of what happens in the universe, and no doubt we'd find that we share a conceptual understanding of it.

At the time when we were trying to understand heat, there were competing theories about it. Some thought when heat gets transferred from one object to another, there's a liquid that gets poured across into it somehow. There were also other ideas, but increased particle motion within a hot object *won*.

It turned out to be the right answer, and yes, there is one. Although it's not a complete description, and we may never reach a complete one (we just get nearer and nearer one), we have a correct picture.

And quantum mechanics, just like heat, will also probably turn out to have a correct picture, and something 'going on' underneath it. At present there are different interpretations, as there were with heat.

But with heat, the picture we eventually reached has a universal quality to it. Although less than a complete description, it's easily accurate enough for us to agree on it with creatures from other planets. So if we all wanted to, once a translating system had been sorted out (which might be a bit of a task), we could sit down with them, and talk about conceptual pictures, even if their mathematics is totally different from ours.

This is more than a trivial point - it's about establishing what's real, and what runs deep. The aim here, rather than to talk about aliens, is to show that the conceptual side is, and has always been, the true backbone of physics. Even if we take it for granted, it's like a structural beam, and can't be removed.

The conceptual side may also be a part of the goal of physics, as Einstein and Wheeler thought. We don't know what the goal is yet. We'll find out what it was all along when we get there. But the ultimate goal may be a conceptual *and* mathematical description of the workings of the universe. If so, the final theory, if anyone had the questionable taste to print it on a t-shirt, as some have said could be done, might need more than just some algebraic text, and an equation. There might be a need for a picture file as well.

43. How real is a positioning of the dimensions?

Talking about what's real, I should describe the picture that DQM provides in a bit more detail. There are questions about what's real in it, and what isn't. One thing worth remembering is that if something is emergent, that doesn't stop it from being real.

An emergent phenomenon is something that comes out of something else, at a more superficial level. If a hillside is eroded by wind and rain over time, the resulting shape of the hillside is emergent. It appears out of many events that combine. The rock itself, in the hill, is just as fundamental as the wind, rain and rivers that affect it. But the *shapes* the elements carve out, as they slowly sculpt the land into hills and valleys, are emergent.

You sometimes get a series of layers, each one emerging from the one below it. And more complicated patterns tend to emerge out of simpler things that are going on underneath them.

So returning to DQM, you have a picture of a relational world that emerges out of something more fundamental. But there's a need to say a little more than just that there's an emergent level.

When you get right down to it, what's really out there is a set of dimensions with no fixed positions. At the very deepest level, the dimensions have no positions at all. This is not only part of the DQM picture, it's also part of the standard picture of the dimensions. But by 'no positions', I mean no absolute positions. In DQM they can have relative ones, and they do that a lot.

In the standard picture of the dimensions, it's the same. There they also have no absolute positions, but they can have relative ones. And they also do that a lot, but in the standard picture, that just means sitting there at right angles to each other.

In DQM, there's more going on. The dimensions are vibrating, there's a single structure, and what we call light is waves of all sizes that go rippling through that structure. Matter is circular vibrations in its fabric, travelling around small cylinders that are part of that structure. And because matter can move, this is enough to create a set of possibilities for how matter might be related to other matter locally, in three-dimensional space. These possibilities take the form of a set of local positionings for the axes. It's important that matter can move around in space, because this creates a set of frames, that is, a set of viewpoints on the relationships between moving objects.

These relationships are what we call entanglements, and they're created by interactions. And they affect what happens directly. They're more than just implied spatial relationships, because they have physical consequences. And what happens is, as soon as a positioning for the dimensions is established, then locally, it's no contest. One positioning then becomes the relevant one, and everything builds itself up around it locally. This creates an intermediate level of structure: it's emergent, and it make our world. It emerges out of many physical relationships between bits of matter. Or to put it another way, it emerges out of many entanglements.

So there's a level at which the dimensions do have implied local positionings. It's not a level of reality, though it sometimes behaves like one. It's a level of structure in the universe. It exists with or without observers looking at it, and in some practical ways it's what we often mean by the word 'reality'. But it's about relational information only, and it differs from the bedrock level in the way in which something relative differs from something absolute.

The emergent level is also, in some sense, built up in three space dimensions, while the underlying level is effectively in four, because the cylinders add an extra one. And strangely enough, comparing the two levels, in terms of the positionings for the dimensions, what exists at the deeper absolute level is 'unfixed', while what exists at the shallower relational level is 'fixed'. This is a reversal of what one might expect.

And perhaps rather typically of products and services, the emergent level is a bit ropey in places - it doesn't always join up too well, and bits of it are not sure where the rest of it is. But luckily that's only at a small scale. At a larger scale, structure builds up, and things hold together. So although the bedrock level is unfixed dimensions, there's a very reasonable level above that. When we observe the wavelike behaviour of light or matter, we're looking at the deeper level. But when we find particle behaviour, we're looking at this more superficial level.

And if one bit of matter is moving in a particular way, another bit of matter might approach it, interact with it, and establish the relationship between it and the first bit of matter. The wave function - the unformed pre-state - then disappears, and both bits of matter will start to behave as if they have a clear relationship. And if there's other matter nearby, and there probably is, in this picture you'll get a wider framework building up, with bits of matter rapidly being slotted into place, rather like Lego pieces.

It's not just bits of matter, interactions often involve photons as well. A large part of what happens is that the photons fly through places where there are bits of matter hanging around, carrying information with them.

As matter gets connected in this way, the framework that builds up is purely relative, and contains nothing absolute. So it might seem less real than what it emerges from. But it's not, it's just less fundamental. Still, it makes what can seem like a separate level of reality. And our species, and other species we're friends with, have evolved taking that level to be reality.

These positionings for the axes that get established are related to reference frames, which are differently moving viewpoints. And like frames they exist, in some sense, with or without observers. In fact, it's the same kind of setup as that of frames, but it's more real than frames are thought to be. It's as if that kind of setup has been promoted up to an extra level of reality, as it's what our world happens to emerge from.

44. Decoherence as in DQM

To me the most interesting thing about decoherence was the way it chimed with something Rovelli said: that matter is always making 'measurements' on other matter, by interacting with it, and getting relational information back.

So splicing these two ideas together, when matter in the wave state interacts with its environment in decoherence, creating entanglements with other bits of matter, and so reducing the number of possibilities in the wave function, perhaps what's really happening is - it's getting *relational information* from the matter surrounding it.

Bearing that in mind and returning to my picture, in DQM, 'getting relational information back' translates loosely to 'finding out where the dimensions are positioned locally'. So in DQM what really happens is that while the wave function breaks down, or decoheres, as the range of possibilities it contains is reduced, something else is building up, and coming together.

And that something is the local framework. Matter joins onto it rapidly, and the way I picture it, it looks a bit like when iron filings fly onto a magnet - my partner Jill suggested this analogy. Matter doesn't move anywhere, it just in some sense 'angles itself' in space at the Planck scale, having got its bearings, by finding out how the cylinder it sits on is positioned. But this positioning of the cylinder is a purely relative one - it's in relation to other matter on other cylinders.

But what makes the analogy better is that iron filings also angle themselves. I can't show this happening at the Planck scale, but if you get some iron filings, a piece of paper and a magnet (hopefully without having to pay for them), and put them on the paper with the magnet underneath, they'll all suddenly angle themselves, and point in a particular direction. This might convey some of the sense of it.

So anyway, we think of decoherence as a reduction. Our present picture of it involves something rapidly breaking down, as a result of many interactions. And what breaks down is the superposition, or the wave state. In my picture, this does happen, but during the same very short period of time a framework is appearing. So the full picture is of one thing breaking down, while another new thing builds up, and the two processes are directly connected.

45. Another clue seen only with hindsight

With hindsight, there was another clue that suggests this solution, although I can't say I saw it beforehand. It involves gathering up some ideas, including a few that have been set out loosely in the last chapter.

If you take two developments in quantum theory from the late 20th century, decoherence and RQM, you find that they both point in the same direction. It's not just that they both talk about interactions. But they do: according to Rovelli's view of RQM, an interaction brings with it an exchange of relational information. And in decoherence, interactions with other nearby matter lead to the superposition being eroded somehow.

These two approaches may have got to the idea of interactions in different ways. Decoherence originally took a mathematical route - it was found that decoherence was already implied in basic quantum theory. But Rovelli was reinterpreting the theory, and developing a new way of seeing it.

He was influenced by Zurek's work - he mentions it in his papers - but I don't know if he got to his ideas about interactions partly from there. My guess is that he did (he told me that by the '90s the interactions view was looking clearer generally) and if so, and if DQM is right, then there lies another part of the trail that got us there. Zurek showed both myself and Rovelli where to look, and his contribution, from decades of work on decoherence, is a major one. He's the main person, along with others who worked on decoherence, who really cracked into quantum mechanics enough to get a foot in the door, and let the conceptual thinkers get in.

Only hindsight will show whether or not this turns out to be the point where we breached the wall, and cracked into quantum mechanics. We'll only know later. But whether or not we did, physics is ultimately teamwork - it's very much a group activity. No breakthrough is ever made nowadays without the earlier contributions of many physicists, and each breakthrough is then taken further by the later contributions of many physicists.

Anyway, however he got there in the context of RQM, Rovelli saw, as others did at the time, that interactions might be the cause. To measure something, you have to make something else interact with it.

Returning to the clue seen with hindsight, it comes, as clues do, from splicing two things together. In decoherence, interactions reduce the undefinedness in the wave state. And in RQM, interactions create exchanges of relational information. Put those two together, you get: *matter's state becomes more clearly defined when it establishes its spatial relationship with other matter*. To me, because my picture of matter on the dimensions was already in place, that should have told me something. It should have said that whatever was happening, a positioning for the dimensions was being chosen.

But for a long time the different bits of that clue were all in different places among many half-formed ideas, some of them more helpful than others, that were only seen more clearly later.

Part 9. Using this picture to interpret what we observe
46. Counterintuitive phenomena

In physics we have a lot of couterintuitive things to deal with. It's not just in quantum theory. Something that's counterintuitive doesn't necessarily need an explanation at all. The world might just *be like that*. So physics isn't about trying to accommodate our intuition. For one thing, we know our intuition may have evolved with a limited view on what's out there.

So an intuition-clinging physicist is, generally speaking, not a good one. There are some, mostly out at the fringes, who allow their intuition to decide what they believe. Some are prepared to ignore or question experimental results, and overlook a large body of evidence. Some refuse to take standard theory onboard because it's counterintuitive, and want to return to classical physics, where the ideas were more understandable.

But nowadays we know from experience that our intuition is unreliable, and should not be depended on - in physics. However this can be misunderstood. The fact that a counterintuitive phenomenon doesn't necessarily *need* an intuitive explanation, certainly doesn't mean it can't have one. It might have one anyway. So if a physicist finds an intuitive explanation for something, it doesn't necessarily mean he or she is an intuition-clinging physicist.

About a century ago the ether was disproved, and it was a shock. The ether, which had been thought to behave like matter, turned out not to exist. Then people decided that light has no transmitting medium at all. Somewhat 'on the rebound' from ideas resembling the ether, people threw out more than the bathwater, and swung right away from the idea of there being any kind of transmitting medium - even a very different one. They also became a little over zealous, perhaps, in the new suspicion of intuitive explanations.

And we found that light could be described as a wavelike disturbance in the electromagnetic field. That's a specific idea mathematically, but an unspecific idea conceptually. We don't really know what fields *are*. Einstein wrote in 1924 that he thought they were connected with 'states of space', and many have thought something similar. In 1947 George Gamow wrote that most physicists thought space itself is what vibrates when light travels. Feynman, in the 1960s, called a field a 'condition of space'.

But it has been hard to connect fields to space in a specific way. This may be because we need to know more about the structure of space before we can do that. But we know that a field can transmit waves. And these waves use mathematics similar to the mathematics we developed for waves on Earth, which have some kind of elastic medium to travel through.

And waves on Earth travel at a characteristic speed for the medium, which is the same for waves of all sizes. This speed depends on the properties of the medium, and how fast it happens to spring back after being displaced. The faster it springs back, the faster the transmission speed. This would fit well with light having a fixed speed, the same for waves of all sizes. It means c might be the transmission speed of space itself. If so, space springs back very quickly. This supports the idea that space can vibrate, and transmit waves, and hence that dimensions can too, as they do in DQM.

But the thing is, and this is what I realised in the '90s, nowadays it's widely believed there's a fine grain of curled-up dimensions filling the vacuum, at a very small scale. So if we were looking for a transmitting medium, and one that might explain the electromagnetic field more specifically, nowadays we don't have to invent one 'out of thin air', as we did with the ether. There's one that we tend to think exists anyway, for entirely different reasons, sitting there already, and filling all of space.

And if that structure really is the transmitting medium for light and matter, as in DQM, then given what it consists of - folded-up dimensions - it might be a weird medium. And a weird medium might lead to explanations for some counterintuitive things in physics. It's not that an explanation is necessarily *needed* for those. But one might exist, just anyway.

And to put it in a loose way: if there's a weird transmitting medium, it would help remove weirdness from elsewhere in the picture. There's often a trade-off. The weirdness has to go somewhere, but it can be kept to a limited area, because a lot gets explained by the properties of this transmitting medium. So that was one of the starting points for the whole theory in the mid 1990s, and it led to explanations for a lot of things.

47. Interpreting the double slit experiment

Having set out the basic DQM picture, next it must be used to interpret some aspects of quantum mechanics. The double slit experiment covers a lot, and has to be dealt with. But there's also entanglement, and I'll give what I hope is a clear interpretation for that. And Schrödinger's cat is another distillation of some key questions, and can't be left out.

Why does light travel as a wave, but depart and arrive as a particle? Between the emission point and the arrival point, while on its journey, why do we get what John Wheeler called the 'great smoky dragon', with its body covered in smoke, representing the unclear states of light (and matter)?

When a photon is emitted, we know where its path starts out. In the DQM picture, it's then immediately moving along a dimension that hasn't yet had an angle in space allocated to it. So from our point of view, it moves along all the paths that could be the relevant path at once, at different angles. The result is a kind of wave. The photon then spreads out in space, making what Wheeler saw as the smoke. At the other end, an interaction picks out one of the possible paths, partly at random.

In the double slit experiment, the path that the light takes seems ambiguous. It can even be decided later, apparently acting backwards in time, in 'delayed choice' versions of the experiment. According to DQM, there's no retroactive decision, because all the different routes were actually taken. So there's just the viewpoint from which we see the experiment. Fixing the dimensions also fixes the light's journey through the experimental setup, picking one journey out of many journeys, all of which it actually took at the same time.

The different versions of the situation all exist because the dimensions are in many positions at once, and everything emerges from that undecided smoky cloud. So the light 'travels as a wave, but departs and arrives as a particle'. Nothing about the past is altered, but there's an intermediate selection level, if you like, which is about fixing some background specifications within which the experiment happens.

This can explain another aspect of the double slit experiment as well. It looks as if putting a detector in the photon's path makes it choose which route to take. But instead a measurement of this kind creates an interaction between the photon and a particle in the lab, and that creates relational information that allows the photon to get its bearings.

And once it's clear what the relationship with the lab is, that decides certain things. If the photon is detected passing through one of the slits, then the dimensional axis along which it moves is angled so it passes through that slit. So one of the axes is, at least briefly, at that particular angle. The selection is partly random: one path is picked out from many possible paths contained in the wave. But it has to be a possible one, and if the photon *does* arrive at the screen at the back, then all the possible paths must go through either one slit or the other. So once a positioning for the dimensions is established, even if very briefly, it goes with a route. The interaction needed for a measurement defines the light's relationship to the wider experimental setup, and in that process, its route through the setup becomes clearly defined.

Just as an aside, there's a funny thing about this. To us, we think we're using the detector to make a measurement on the photon. But in a way, it's more a case of the photon using the detector to make a measurement on us. Well not on us, but on the lab we managed to get a job working in.

I've seen a group of Californians experimenting with a group of dolphins, and playing visual games with them. Inflatable shapes get associated with other inflatable shapes, and one object has to be put with another, at the corners of the swimming pool. The dolphins seemed to understand this, but they did unexpected things, that had a different kind of logic to them - perhaps their patterns were just as rational as ours. It almost (but not quite) made you wonder if the dolphins thought they were experimenting on the Californians. This reversibility of the situation has no real parallel with what happens with a photon, but there's a grain of something to it. Although the photon doesn't have a viewpoint as we do, you might say the detector tells the photon more about us, than it tells us about the photon.

But to interpret, and hopefully make sense of, the double slit experiment in a more complete way, there's something else. Before we collapse the wave function by putting a detector near the slits (to get particle behaviour), the wave already seems to contain particles, or the particles already know about a wave somehow. This is the 'one particle at a time' version of the double slit experiment, where we still get an interference pattern building up.

According to DQM, what we're seeing when this pattern builds up is really a composite wave, made up of separate parts of many different waves. When particles are sent through the experiment one at a time, whether photons, electrons, or whatever else, *each of them* spreads out into a wave because there are many positions for the dimensions. Then each wave is collapsed to a single trajectory by the interaction at the screen at the back. That happens via a partly random selection process.

But when many of the trajectories are picked out over time, and put on the screen together, one finds they still make a wave. It's not a single wave as it has often been taken to be, it's many waves that all existed separately, in the same place, each with their own individual interference patterns. But all the trajectories, each from a different wave, when they're combined, still make an interference pattern. Superimpose bits of many interference patterns, and what you get is - an interference pattern. It arises from pieces of many such patterns, and it looks like there's only one wave.

And the fact that the wave can be seen being 'reconstructed' out of particle trajectories actually supports one of the very central premises of DQM: that a quantum wave is made up of many separate trajectories for particles.

48. Schrödinger's cat

I've done almost everything except entanglement now, but I can't leave out Schrödinger's cat. The internet has endless discussions, some quite confused, about the life and death issues the poor minou faces. And the paradox pulls together some key questions.

To me the main question is about the superposition. How literally do we take it? Is it a real physical superposition, a configuration of information, or both? Schrödinger's cat is less about the large-scale cat, and more about trying to hold a powerful magnifying glass to the small-scale superposition, assuming we can amplify it, and asking what would happen if we did.

But that's not what Schrödinger was doing. He was pointing out what he saw as the absurdity of one result of the current view. In the 1930s the prevailing interpretation, Copenhagen, suggested the cat would be both alive and dead at once. He was challenging people to find a more reasonable interpretation, though ironically, some later views included what at the time seemed too absurd to consider (that actually happened with EPR as well). But anyway, in later years, as the interpretations spread wider, the cat remained a central question, because it was above all about how to interpret the theory.

I won't outline the setup, it can be found easily enough. Schrödinger was an excellent physicist, but perhaps not a 'cat person'. The small-scale part of the experiment involves the random decay, or not, of a particle, which is used to create a 50 - 50 situation.

That much may be in a superposition, as it's thought to be. We can't check it directly. We normally identify a superposition via the interference patterns, and in this case there isn't one. But although some setups thought to create superpositions are questionable in DQM, let's assume there's no problem with the mechanism that sets the whole conundrum in motion.

The next question is whether that small-scale superposition can then be sent upwards, extrapolating up and away into the large-scale world, by attaching it to a situation. Again, the large-scale superposition is hard to confirm. If we choose to open the box and look, it disappears - that is, if there was one. We can assume that there was one a moment ago.

But if we do, there are issues about scale. We'll return to the cat in a minute, but there's some background to fill in first. We've found that at two different scales the world behaves very differently. The wave state doesn't really exist at a large scale. It disappears in a small fraction of a second. Objects that we can put in a superposition in the lab have got steadily larger, but they're still small. They'll probably get even larger in the future, but that's an increasingly artificial situation.

During the first half of the 20th century, many thought there were two sets of rules, one for each of the two scales. Nowadays we tend to think there's just one set of rules, and the world is quantum at all scales. So to me it seemed that we probably need to find something *emergent* to explain the large-scale behaviour. That's how you get a transition, but keeping the underlying rules.

When I got to that idea, I assumed that others would also see it that way, but instead I found people saying all kinds of things. One respectable view is that as we have no explanation the large-scale world's different behaviour, or for how it crystallises out of the fuzziness of the small-scale world - it's not clear if quantum theory can be applied to the large-scale world at all.

But the scale issue is one of the areas where we've made progress since the original cat problem was pointed out. Decoherence has shown clearly that the relationship between scale and the *speed* at which the change happens is real. The larger the object, the faster it jumps out of a superposition. This is the first handle we've had on the scale issue, and it shows progress is being made.

And not only have the predictions of quantum mechanics been confirmed by experiment, it has been possible to watch the reduction in the coherence of the superposition happen over time. It has even been possible, for instance, to control the coupling between a trapped ion and its nearby environment, making sure that what decoherence is about - interaction with the object's environment - is actually happening.

According to DQM, the large-scale environment includes something that can be latched onto. At the scale where we live, there's a 'local framework'. The dimensions have implied positions, bits of matter have relationships, and a lot of nearby matter has already got its bearings. So you simply can't have an object there in a superposition for long. It immediately interacts with nearby light or matter, and joins onto the framework. That then picks out one of the possibilities (which are possible orientations). Or in the language of quantum theory, the superposition decoheres. This takes a very short period of time: it varies, but an example is 10^{-17} seconds.

So in DQM, this recently discovered relationship between scale and speed is explained because in the large-scale environment, there have been a load of interactions. They cross-corroborate each other, and the relationships all tell the same story. The larger the scale, the more information about the existing setup is available, *so the quicker that kind of information will go everywhere locally*. That is, assuming you have matter all around to interact with, telling the undecided matter in the experiment what's where.

Returning to the cat, the larger the object in a superposition, the quicker it interacts with nearby matter, and the faster the superposition disappears. So any large-scale superposition, such as a big one the size of a cat, rapidly goes to one state. The cat doesn't have time to have a near-death experience, or to notice anything odd. The randomness quickly changes to just an either/or kind of randomness. And very soon the cat isn't dead and alive at the same time, it's simply one or the other, like any other cat.

So Schrödinger's cat is like throwing a drop of water into a fire, or a spark of fire into water. Yes, it may briefly exist like that, in a strange state, like a tiny pocket of one thing sitting in another. But it doesn't last long.

(And strictly speaking, we now know about decoherence: although the vial of poison may break, it will have no time to take effect. So the cat will briefly be both alive and in grave danger, and also alive and not in grave danger, at the same time. That's the only large-scale superposition, if there is one at all. It will then be either alive or dead, but not both.)

But this whole view of the question comes from assuming we can 'extend' the small-scale superposition, and amplify it. What this actually means is, we assume that individual possibilities within the superposition are real enough to set off chains of events that use cause and effect. That's what Schrödinger envisaged (Geiger counter tube, relay, hammer, flask). All of that was within just one possibility, out of more than one. Perhaps that can happen. If it can, we can split the world into two separate worlds.

It's not just the cat: although decoherence puts powerful limits on our ability to do anything much with this, it's about dividing the world into two worlds. And hypothetically, although it gets very, very speculative, if we could shield matter from its environment enough to protect the superposition for longer, we could create two ongoing stories, with preprepared chain reactions (the live cat gets a bowl of crunchies, which sets off a series of other things). Both alternative sequences would exist at the same time: two parallel worlds.

What does this tell us? Not much really. It's not the same as the many worlds interpretation, but perhaps the fact that we simply can't split the world into two worlds (because of decoherence) adds slightly more reason to think the many worlds interpretation is wrong. The world we find doesn't seem to like being split into two, with the two different versions extremely close together in one particular place. It collapses back to one world in about 10^{-17} seconds, and preventing that may be impossible.

And if a given superposition ultimately arises from different positions for the dimensions, as in DQM, then I expect it would be very difficult to 'extend' the superposition to create a series of events. For one thing, to make a world at all, you have to make a lot of complicated relationships in three dimensions. But doing that is what collapses it back to one world.

49. Entanglement

There's a very specific setup in quantum entanglement, and it fits this picture particularly well. It also tells us something about how the universe is, if DQM is right.

But first, there's a simpler clue, about how entanglements get started. The clue strongly supports the interactions approach, and so DQM as well. There are two well tested ways in which we know entanglements can be created, so again, this clue comes from splicing two ideas together.

Firstly, two particles can get entangled as a result of *starting out together in the same place*. Entanglements between pairs of particles can be created in the lab, which is done at the initial stage, when they're together. An earlier particle splitting in two can leave two entangled particles behind it.

And secondly, in decoherence, we know *interactions create entanglements*. The series of interactions between matter and the matter surrounding it, will create entanglements between them. That's what decoherence is about, and it's very strongly supported - both by direct experiment, and also by coming from a theory that has never been wrong.

Put those two together, you find that both processes lead to entanglement, and in both cases, the particles touch each other first, in one way or another. Both involve physical contact, so what you get is: *physical contact between particles creates entanglements.*

So it seems that starting off in the same place is equivalent to a collision. This may be because in both cases, the particles have been connected physically. As a result, they somehow have a lasting relationship. Entanglement is about particles that have already established a relationship somehow, and it seems that physical contact is how they do that.

That clue, about how entanglements get started, I'd say is a major one. Then there's also the more familiar set of clues about entanglement. You can have a pair of particles, and even if they're separated by a large distance, if one of them then has a measurement made on it, and is found to have a certain property, the other can *instantly* take on another related property, and from then on will possess that property.

Say a photon is sent through a setup that splits it into two photons. The new photons can be made to have perpendicular polarization directions, so one of them can be taken as horizontal, and the other as vertical. But quantum mechanics doesn't tell us which is which, *or where that right angle is*. Before a measurement is made on one of them, neither particle has these kind of properties at all, except as a kind of weird set of conditional possibilities.

We've known about this since 1935, and it has been a major mystery. It has also been a cause for a fair amount of concern. This is because it suggests an effect that works instantaneously. If such an effect exists, then it might break the speed limit c in relativity, which to some is a speed limit for any influence to travel. So some see it as a threat to the consistency of standard physics. But we know it happens, pretty much, coming both from confirmed theory, and recently from direct experiment.

Looking at the clue, it's very specific. We know we can have pairs of particles with no actual properties, but with the condition that 'if x is true of one, then y is true of the other'. Now not only is this a very specific clue, it's also an exceptional one. By that I mean there are a lot of possible pictures of what's going on that simply wouldn't fit with it. And that includes just about all the pictures we can think of, because none of them do. So if a picture is found that genuinely does fit with it, it might be the right one.

In DQM, the condition of being entangled is the condition of having to have the same positioning for the dimensions. The two particles have touched, whether from a collision, or from having started out physically connected. So they've either had an 'exchange of relational information', or they were born with the relational information in the first place. Either way, it defines their relationship, as it links them up via an implied positioning for the axes.

So when we know something to be true about them, for instance: 'If particle A has the property x, then particle B has the property y', what's really meant is 'If the dimensions are *here*, then particle A has the property x, and particle B has the property y. But it might also be that the dimensions are *here*, in which case it's the opposite. And if the dimensions have no implied positions in relation to us at present - us in the lab, that is - then the pair are for now left hanging in mid-air, and we can only say, well, they're linked in such a way that 'if this, then that, and if that, then this'.

This reminds me, again, of doing a jigsaw - sometimes chunks of jigsaw don't join up (this is not the puzzle solving analogy I sometimes use, it's about a real jigsaw this time). You can have a chunk of jigsaw like an island, placed vaguely out on its own somewhere off the mainland, and we only know for sure about the connections within the chunk. But you often get something of an archipelago, with many little islands, and sometimes we have some idea of how one will relate to another.

Something that's rather like entanglement can actually happen when doing a jigsaw. Two small chunks of jigsaw, or two loose pieces, can have an implied connection across a distance. For instance, the two ends of the balcony of a building might each be on a separate piece, and both pieces might be found and identified. But until we know where the building goes, and how the balcony is angled on the building, all we know is that if one piece goes like *that*, then the other piece must go somewhere like *that*.

Perhaps, looking at this, you can see the way that bits of the universe have implied relationships with other bits of the universe in this picture. There are these separate incomplete elements of the world, each with their own little story implied, and all disconnected from each other. And looking at how our world constructs itself, even when the connection between two particles is strung across a very large distance, the 'little story' can still be there. But it can be a *potential* for something that awaits a decision, rather than anything complete. So the universe is capable of holding onto these partially complete pictures, and keeping the potential relationships between bits of matter as they are - perhaps for billions of years.

But when the situation finally gets resolved, and one of the pair of entangled particles bumps into something else, it doesn't join onto the mainland. There is no mainland. There are only smaller islands and larger ones. So a pair of entangled particles is a small local framework, while a laboratory is a larger one. In both cases, at least bit of connecting has been going on. (The word 'local' is used because these frameworks usually are, but the connections can actually stretch across large distances).

There's no mainland simply because if there was, that would be something absolute, a bit like a preferred frame of reference. But these connections are relational only, which is why there's no centre of things.

In order to show the limits of the jigsaw analogy, I should say that the aim of the game is not for the universe to put all the pieces together and eventually make a complete world. Instead, the 'smaller islands and larger ones' are like reference frames - local viewpoints on the universe - and they co-exist in the same way that different frames do.

Having defined entanglement in DQM's terms, I can remind you of a point I made earlier in this chapter. In decoherence, when an object interacts with its environment, it seems to get connected up with it somehow. When that happens, what's actually created is entanglements between bits of matter in the object, and other bits of matter surrounding it. The existence of these entanglements, or their creation, somehow reduces the number of possible states in the wave function, and the object then - for some reason - becomes more clearly defined.

The DQM explanation for entanglement, as *spatial relationships* based on shared positionings for the axes, works very well on its own. But it also works as a way to explain this known element of decoherence, with interactions establishing the spatial relationships. And in general, if you substitute for the phrase 'quantum entanglements' another more cumbersome phrase, 'spatial orientation relationships', you do get clumsy long sentences, but good sense comes out as well.

The only thing left to tell you about, for Book I anyway, is the instantaneous aspect of entanglement. Quantum theory tells us the effect is instantaneous, and the direct evidence we have for that has moved steadily closer to going beyond any doubt, in an increasingly accurate series of experiments.

All that has been proved so far is that it's way faster than lightspeed, but the theory has been right so many times, most of us think these instantaneous connections exist somehow. The apparent contradiction between that and relativity theory has been seen by some as the deepest crack in the standard picture, and it would be good if the dimensional interpretation for quantum mechanics could shed some light on the question.

50. Instantaneous connections

When non-local connections were first identified by Einstein and friends in 1935, their aim was to use them to show that quantum mechanics had to be wrong or incomplete. 'Spooky action at a distance' seemed impossible, partly because it seemed to go against the speed limit in special relativity. So their argument at the time was that quantum mechanics simply can't be like this, because if it was, then there'd be these ludicrous instantaneous connections. But unexpectedly, these instantaneous connections turned out to exist, and we've been trying to explain them ever since.

EPR (the 1935 paper from Einstein, Podolsky and Rosen) and Bell's theorem taken together, have apparently worrying implications about the nature of reality. They seem to show that non-locality - non-local connections - must exist. The wry physicist David Mermin says there are two types of physicists:

those who are worried by this, and those who aren't. But he says the second group subdivides into two further groups, those who attempt to explain why they're not, and either miss the point or make assertions that can be shown to be false, and those who refuse to explain why they're not. So anyway, as you can see, the problem isn't a minor one.

According to DQM, what we call reality very often means a world with fixed positions for the dimensions. But there's a deeper level sitting underneath it, in which things are unfixed. Out of this unformed raw material, our familiar 'level of reality' emerges, though in fact it's just as real as the deeper level. These two physical levels that exist in the universe are different from each other, and they differ in the way that a map with no compass points on it is different from a map that has just had a fresh set of compass points drawn on it.

Before the pen lands on the map, you might guess that North will be at the top, and it might, but that's a convention. North could be anywhere, so until someone draws the compass points, we don't know. But once they've been added in, the compass points will immediately affect the relationships of an enormous number of bits of land and coastline across the map. This happens instantly - their influence doesn't spread outwards at the speed of light, or at the speed of waves in map paper, or at any other speed.

Compass points are like a co-ordinate system, and so are the dimensions. A co-ordinate system is a reference system that accompanies the territory to which it refers. It provides 'relational information' - when the compass points are drawn onto the map, we suddenly know that John o' Groats is north of Land's End, that Brazil is west of Mozambique, and that one grain of sand on a beach in France is south-east of another.

But to make the analogy more accurate, we're looking at a very small area of a huge and frankly rather badly administered map, on which different areas can have the compass points in different positions, and they sometimes need to be established or re-established. So returning to the puzzle, if a particle interacts with a measuring device in a lab, it will find out where the compass points are, and get its bearings. The lab, being at a large scale, already has a local framework in place, which comes with the building, whenever you rent some laboratory space.

Now let's say you have a pair of entangled particles. Their relationship means that they have preconditions and correlations about certain things. In DQM these are really about the axes, and where they might be. Then you send one particle a very large distance away, too far for a lightspeed signal to connect them over the time frame of the experiment. Then you make a measurement on the particle that's nearby, making it interact with matter from the lab.

The question is simply: why will the far distant particle be affected instantly, as quantum theory implies, and as experiment has suggested?

To make the measurement, you have to make the nearby particle bump into light or matter that has already got a fix on the laboratory frame, and carries information about where the dimensions are locally. Whatever information comes from the interaction that allowed the measurement, it will also affect the other particle, which is linked to the first, even at a far distance. But the effect it has will be in relation to the lab frame, rather than absolute.

The reason that the distant particle will be affected is this: what two particles being entangled really means is that having gone through what they've been through together, they have a spatial relationship. It means that they both have to have the same positions for the dimensions.

That precondition already exists. It's what I meant by the 'little story' that's involved. It seems that if relational information is created, that has a lasting effect. But it can be incomplete relational information until someone comes along and adds to it, by joining it onto the rest of the picture, or rather, onto a larger chunk of the picture, such as a laboratory.

So putting that in the earlier analogy, suppose the jigsaw piece with one end of the balcony finally gets slotted into position on the mainland (the jigsaw mainland). It's then *instantly* known where the other piece should go, even if the other piece is floating out in space somewhere, and no-one has built the balcony yet. The two jigsaw pieces have a spatially based relationship that already exists, with or without a context. In the DQM picture, one thing the universe can do is similar to this very specific setup.

This explanation for quantum entanglement doesn't compromise the speed limit for light and matter in special relativity, because it isn't light or matter that travels instantaneously. It's something to do with the dimensions. And although that might be called an 'influence', and influences are not meant to exceed the speed limit, it's not a 'light and matter' one. The instantaneous effect is something else, and it's exempt. It's different from the kind of things that have to go by the speed limit, and that set of rules. There's more about this in the next chapter.

But what *actually makes an instantaneous effect* across large distances? The answer is, the two entangled particles, in the simplest version of things, are now two places where the same cylinder is vibrating. They're both on one of the cylinders. And any one of those cylinders has length, because it includes a flat dimension. This way of putting it is an oversimplification, but once the relationship between the two particles has become clear (for instance if they started off in the same place, or interacted at some point), from then on this cylinder stretches all the way from one particle to the other. It has become clear where it is, and where they are on it. So the cylinder links certain things about them across any distance, however far apart they may be.

The properties of the dimensions allow this to happen, given that matter is disturbances in their fabric. So even if it's not clear, for the time being, how that cylinder is angled in relation to the matter in some laboratory on some odd planet somewhere, the two particles are still both on the cylinder, and each of them is a part of it.

Part 10. Final points, and a little speculation

51. Removing some possible errors

DQM is also called the dimensional interpretation for quantum mechanics. It has non-local connections, which are explained via a physical mechanism. So DQM is a non-local theory.

Bell's theorem, in the 1960s, ruled out a group of theories called local hidden variables theories, and showed they couldn't fit with quantum mechanics. These local theories include what's known as 'local realism'. But DQM, being a non-local theory, isn't among them.

More recently, the surrounding landscape has got more complicated, and it matters how one defines the word 'realism'. There are perhaps five different definitions of realism. According to some of them DQM is a realist theory, according to others, it isn't. A well respected physicist called Travis Norsen published a paper about a study he did on the meaning of the word realism, as in the phrase 'local realism'. His final conclusion, which he sets out clearly at the beginning, is that the word should be banned. He decides it has no consistent meaning.

A basic definition for realism, as it's used in physics is: *"The idea that nature exists independently of man's mind: that even if the result of a possible measurement does not exist before the act of measuring it, that does not mean it is a creation of the mind of the observer."*

In those terms DQM is a realist theory. In my view there's a world out there, that exists with or without us, and quantum mechanics is nothing directly to do with the mind. Another definition for realism is: *"the view that reality exists with definite properties even when not being observed"*.

That still fits with DQM. It's a bit more complicated, because I think *some* observed properties, not all, depend on the viewpoint. But that's known to be true in special relativity, so it's hardly controversial. And even perspective effects on Earth come near to being properties of an object that depend on the viewpoint. But I nevertheless still think there's a world out there, with *some* definite properties - plenty of them - when no-one's looking at it.

Some people have tried to generalise about theories with realism, but they used a different definition of realism. And my DQM and Rovelli's RQM, for instance, don't have that kind of realism. The definition they used was the idea "...*that **all** measurement outcomes depend on pre-existing properties of objects that are independent of the measurement.*" What about relativistic energy? Let's leave that one.... Anyway, because RQM and DQM both involve matter having **some** frame-dependent properties, this doesn't apply to them. What's more, the definition above says 'of the measurement', rather than 'of the observer'. I'd say it's a bad idea to try to categorise theories in this kind of way.

DQM is hard to pigeon hole. It might seem similar to some hidden variables theories, but it actually has far more similarity to objective collapse theories, and even more to RQM. It differs from hidden variables theories in important ways. For one thing, it doesn't say that quantum theory is incomplete. And it doesn't conflict with there being a genuinely random element, which can't be predicted. Hidden variables theories are very often attempts to avoid that randomness, by saying that there's something non-random going on, to be uncovered. DQM doesn't do that, but it does add concepts.

DQM doesn't have a problem with non-local connections breaking the speed limit *c*, which comes from special relativity. Instead it takes standard physics slightly differently in one particular area, and is perfectly compatible with special relativity. DQM says the speed limit for light and matter in special relativity is just that - a speed limit for light and matter. Many experiments have shown it to be that, but they've shown nothing else. At the time the theory got established, light and matter were all that was known about, so *c* was made a speed limit for everything.

But it's arguable that this was an unnecessary assumption. Above all, it's an untestable one (except by falsifying it). That's because it's an assumption about anything else that might perhaps exist, that we don't yet know about. Why assume anything like that? Why make assumptions about things that science might or might not discover in the next 10,000 years? We're not in a position to do that.

And in fact, after the first 100 years, we're already approaching unavoidable proof that something exists that does indeed break this universal speed limit. And it comes from both theory and experiment. So now, of these two slightly different interpretations of special relativity (speed limit universal, and speed limit for light and matter only), one has been confirmed, while the other is close to being falsified - and arguably has already been falsified, by loophole-free experiments done in 2015, by at least three separate teams.

In DQM theory, *c* is a speed limit, but it's not a universal one. The dimensions are a transmitting medium for light and matter, and their nature, and the way in which they're 'connected up', allows them to connect things across the universe in a way that's not related to this kind of speed limit. To assume that about the dimensions, one can use things that are already thought to be true about them anyway. The dimensions are thought to have properties that potentially make this possible. But unless one happens to think that light and matter are vibrations in their fabric, as in DQM, those kinds of properties are generally irrelevant to this question, so they were not seen as significant. On top of that, in the early 20th century, the dimensions were generally not taken literally anyway.

And this particular change to the interpretation of special relativity is a very small one. All you have to say is that the speed limit *c* is a limit only for things that have been shown to go by this limit via measurements, that is, matter and radiation. And arguably science itself, and its underlying principles, such as ones about keeping theories testable, would tend to say the same.

Moving on, I'll quickly try to clear up two possible misunderstandings, that might perhaps crop up at some point. Firstly, DQM is a non-local theory, as it has a physical mechanism for non-local connections. But because it includes something called a 'local framework' (which is an emergent property of a large-scale place such as a laboratory), and because this is an unfamiliar idea, the word 'local' can be misunderstood, if one has not looked into the theory. But it's certainly not a local hidden variables theory, and the word 'local' is being used in an entirely different way, about something else.

The other possible error is about the phrase 'interaction-free measurement'. This can be misleading - in one use of it, the subject matter turned out to be neither interaction-free, nor a measurement, though it was 'detection-free'. (And R. M. Angelo has argued strongly that this kind of view is interpretation-dependent). The Renninger negative result thought experiment shows that the *lack* of something happening can be enough to set off the sudden change in quantum mechanics. Sometimes the lack of a measurement seems to 'tell the universe' to partially collapse the wave function. Renninger's approach is consistent with the theory, so it may be right, and other related experiments have actually been done, and seem to give results consistent with quantum mechanics, as experiments do.

But if so, the fact that the wave function, or the wave, somehow seems to 'know' something holistic about the setup could equally well be due to the lack of an interaction, rather than of a measurement. Given that so far we've had no explanation for either concept in the context of quantum mechanics, they're effectively interchangeable.

So anyway, those are a few points that might falsely appear to detract from the dimensional interpretation for quantum mechanics. I've briefly tried to pre-empt them here, to explain why they don't apply as they might seem to, and to show why they can be misleading.

52. What's inside an atom?

The 'multiple images' picture of a quantum wave (though it's really multiple versions of the same event), is a way to get a handle on the bizarre world of quantum mechanics. It's a conceptual key, which is what's needed.

With it, one can start searching through what we already know, looking for places where things might fall into place, or fit the picture, or not. One place to look is inside an atom. What we've found there makes very little sense to us so far, but the inside of an atom is near to being literally visible now.

The Greeks thought about atoms thousands of years ago, and in the late 19th century we started to get at them in experiments. Early attempts to model what goes on inside an atom, such as Bohr's planetary model, made an atom look like the solar system, with the electron orbiting the nucleus. My father described that to me when I was a boy - I now know the physics he gave me was more than half a century out of date, but it probably helped at the time, just as that model helped in its time.

But Bohr's 1913 model was one of several attempts that failed. There was one from 1902 with little cubes in each atom, and electrons at the corners. Then there was Thomson's 1904 'plum pudding' model, but it failed too - not only was there no proof of the pudding, Rutherford noticed in 1911 that a particle fired at the pudding had bounced back off something very hard. He realised that it wasn't a plum stone, and the nucleus was discovered. By the 1920s, thanks to Schrödinger and Born, we knew that the cloud surrounding the nucleus of a simple hydrogen atom contained one electron that could be in many places, and they had probabilities attached to them. But it seemed impossible to say how these positions for it were connected up.

Since then our picture has become more detailed, and it has been confirmed by increasingly accurate experiments, but it hasn't really changed much. It's still almost certainly incomplete as a conceptual picture of an atom. We have a single electron making a spread-out wave, that fills a particular volume of space in an atom, with varying probabilities for a location that the electron doesn't yet have, but will get if a measurement is made. The wave is known to contain patterns at the scale of the atom, at a somewhat larger scale than that of the electron itself. They're like standing waves on the skin of a drum.

That's the current picture, loosely speaking, but Bohr's false model persists as the intuitive picture - the electron clouds are still called 'orbitals', although we now know that in a simple hydrogen atom, the electron has no angular momentum, so it isn't orbiting. And atoms are depicted (and atomic energy symbolised) in icons with nucleus and orbits.

Anyway, returning to the history of it, during the first few decades of the 20th century, this developing picture of the atom went through a series of quick, exciting changes, when experiment finally connected with the particle scale, and one by one the barriers fell.

Nowadays 'photographs' have been taken of atoms, but they're several steps away from being a direct image of one. It's even possible to 'see' the electron cloud that surrounds the nucleus (although we still don't know what it is), by observing a large number of atoms, and each time collapsing the electron to a single position, and then putting all their positions together. One method involves stripping the electrons off the nuclei, and watching the patterns that they make afterwards.

There are some strong hints that the wave state or cloud in an atom, say in a simple hydrogen atom with one electron, arises *directly* from a single particle somehow, as it does in DQM. For instance, the electrical charge on the wave is the same as that on the particle.

53. What else is inside an atom?

This chapter has speculative areas, because it's about a very recent idea. And one has to be suspicious of ideas, particularly recent ones. They need time to settle in, and careful checking. They also need pursuing and investigating. But I came to a place where the path I was following split into many different paths. The theory then went into so many different areas that there wasn't time to follow all the paths. But this one still seems worth writing down.

I'm going to outline what might just lead to a further stage to this developing picture of the atom, although it's not a hard and fast part of the theory. It's too early to tell where it belongs - it may or may not become a part of DQM. Work on the theory has been slow, there are always new, half finished ideas, but there comes a point when a line needs to be drawn.

It's also at best an oversimplification. But I think I can justify including it here, because it does help to illustrate the 'multiple versions' aspect of the DQM interpretation, of which I'm absolutely certain. And hopefully it makes one think about this interpretation, as I've been doing. So just as Bohr's planetary model helped us on the way in its time, perhaps this picture will help on the way too, even if it's less than accurate, or less than complete.

Anyway, I've put it here, near the end of the book, with that in mind. To see some raw ideas being knocked around, this chapter might be fun. But some will prefer the paper, which is more considered. (It may be online soon as a preprint, and recently got through peer review for a journal - it's called 'An interactions-based interpretation for quantum mechanics'.)

I'm not one of those people who is still hoping to find classical physics hiding in quantum physics or in other theories. Nevertheless, in my picture of things it's possible to be specific about a quantum wave, and to break it up into a single action that is repeated many times.

And there's reason to think, as I mentioned, that the possible positions for the dimensions can be taken to radiate from the nucleus of an atom, from a point at the geometrical centre. If so, the simplest wave will actually be the behaviour of a single electron, repeated many times on many superimposed radial axes, all set at different angles.

To me the interesting question is - what happens on just one axis? It seems clear that there you'll find particle behaviour. That would make good sense, because when a positioning for the dimensions is picked out, at the point of a measurement, the particle nature is what remains. And in the background theory, that involves straightforward motion along a single cylinder.

And it may be that in a hydrogen atom, for instance, the electron is moving rapidly inwards and outwards on a radial path. In that situation radial motion has been detected, but no orbital motion. We often think of the electron's behaviour as being like standing waves. But if these standing waves, which fit the mathematics in certain ways, can be said to travel in a direction, loosely speaking, nowadays they're seen as being positioned in radial directions, rather than being on orbital paths as in Bohr's model.

So there may be some structure in the electron's particle behaviour, which translates to its wave behaviour. It's known that in an atom, the wave makes patterns, and at a larger scale than that of the electron. They behave like standing waves on the skin of a drum, as in the well-known analogy. These patterns can be spherically symmetrical, varying in the radial direction.

This general point tends to support what I'm saying. There seems to be some large-scale (well, larger-scale) coordination and symmetry within the wave's behaviour. That doesn't fit well with any other direct conceptual ideas we have, but it fits very well with the idea that you're getting multiple versions of the same bit of action, along many radial lines.

So one can then think about what's happening on just one radial line, with an electron on it. It's possible that the electron is moving inwards and outwards radially. Rather than offering a cause for that, I'll give an example of one. It's not necessarily even a possible cause. But suppose the inward pulling force is balanced by an effective outward pushing force, but they vary in strength differently with radius. The electron might settle into a pattern of moving rapidly inwards and outwards. That example is just an illustration, and I hope it helps to sketchily fill in one blank area of the picture.

There are different energy levels, for which we have the mathematics. These levels might correspond to different *stable* ranges of radial motion. So if it gains energy, for instance, the electron settles into a different pattern, and travels inwards and outwards across a different radial range. It's known that only certain energy levels are allowed. In this picture that would be because only certain patterns of motion of this kind are stable.

One reason the idea is of interest is this. Because the electron's location has not been pinned down, the axes are positioned along all radii - they land on many radial lines. And because the electron travels along a flat axis, you get many 'images' of it, rising and falling all around the nucleus at the same time. So when it's near the centre on one radial line, it's near the centre all around the nucleus. The little cloud is oscillating rapidly, and it moves in and out like a sheet. We know the exact shapes of the clouds that surround the nucleus, which are about the probabilities for where the electron might be found. But we don't think of them as being related to anything oscillating over time.

But it may be that the probabilities in the wave function for the electron can be decomposed into (or rederived from) two separate sets of probabilities, with different origins. The first component is about how the axes might be positioned, so it's to do with on which radial line the electron will be found. The second is about *where* on that radial line it's likely to be found - at what radial distance - so it's about the action that happens on that axis.

The result is very like a spherical coordinate system. The probabilities for the second component are related to how the electron's speed changes as it travels up and down. That relates to how much time it spends at a particular radial distance - or rather, for the mathematics to work out, on a particular very small section of that line. This means that the electron's location might be partly dependent on the time at which the measurement is made, in the sense that if we knew where the electron was at a specified earlier moment, that could affect the probabilities for its position at a later moment.

I won't go much further into these ideas here, but it's worth saying that they might add structure *within* the probabilities. We know that QM is complete in important ways, in that no external additions are needed. But within the probabilities of the Schrödinger wave function, in certain situations, there might be more detail to be added.

If so, this could be partly time-dependent. If the cloud is made of multiple versions of a single electron that has specific motion, then the probability of finding the electron at a given location would have an extra variation to it. It would depend on where the electron was in space recently, at a given point in time. If it was possible to get extra information about earlier points in time (and it's difficult, but perhaps not impossible), then the probabilities about a later point in time might be affected.

Sometimes the orbital takes a form with separate 'lobes', and you have to try to explain why they don't connect up. Perhaps the electron will simply be found in one of them, and will not travel to the other. If that's the case, and I think it is, then if you don't know which volume of space it's in, the odds about where it might be found are still the standard ones. But if you do, then again, information on where the electron was three seconds ago can have bearing on where it might be now.

When there's more than one electron it gets more complicated, and because they repel, the lobes can distort each other. In my picture, this can also limit the possibilities, and if one electron is on one side of the nucleus, the other might have to be on the other side. But anyway, for the time-dependent side of things, this extra information that's needed - about where the electron was at an earlier moment - is hard to come by. Looking just once collapses the wave. But on the face of it, you need to look twice.

It would be difficult to measure this for a range of reasons, and the fact the Heisenberg's uncertainty principle smears everything out doesn't really help. But people might find a way, probably using more than one atom. This would also relate to the way atoms bond together - they might be synchronised in their oscillations. And electrons are also sometimes shared between nuclei, so they might be taking a route that involves both.

To examine this idea closely, one would need to look into a range of different aspects of it - I don't have time to take it any further at the moment. I hope that myself, or others, or both, will take it further in the future, and it seems that progress might be made by looking further into it.

But whatever comes out of that, a quantum wave should certainly, without any doubt at all, be taken to be made up of the same action repeated many times, at many different spatial orientations.

54. Defining this interpretation

The usual way of classifying interpretations for quantum mechanics doesn't easily accommodate the dimensional interpretation. It's difficult to place it on the standard spectrum, which has a range of different ways of seeing QM. Words like 'reality' turn out to need defining very carefully or redefining, and have new shades of meaning that make DQM hard to categorise.

There's a page on the internet that tries to separate off the different views of quantum theory by asking a series of questions. It filters the different views, and you follow a route across the diagram, arriving at boxes that ask things like 'do you believe quantum theory describes an independent reality?', and how you answer affects your route onwards from there. I found my answer was sometimes 'well... yes and no'. And I felt like adding 'I'm not trying to be difficult, but to answer truthfully, I need to say that'.

To me the diagram oversimplifies and misrepresents things, but to be fair to whoever made it, it's not that easy to say *anything* general about quantum mechanics. There's always the risk of including a false assumption, using the standard ways of seeing it. And according to this picture, the very language we use contains false assumptions built into it, and it's nobody's fault. But as a result, DQM finds its feet naturally in a visual form - it's ultimately a visual solution.

People sometimes list criteria for interpretations of quantum mechanics, and what they should contain. They sometimes include rather a lot of the present landscape, some areas of which might be changed by a new interpretation. This can make it harder for genuinely new ideas to get through, by making them fill in a form that forces them into the clothing of some of the present ideas. On the other hand, Christopher Fuchs says: '*I say no interpretation of quantum mechanics is worth its salt unless it raises as many technical questions as it answers philosophical ones.*' So he seems open to new ideas that might shift the present landscape around.

And this one does. There's even difficulty fitting it into the standard set of definitions. For instance, when one tries to define the word 'reality', one has to refer to the interpretation in the definition. Just as an aside, It's often thought that our reality appears out of something less well-defined, and less real. Some people think quantum mechanics shows the world appearing out of something unreal, and becoming clearly defined.

But according to the dimensional interpretation, our reality emerges out of something that is, if anything, *more* real, not less so - or more fundamental. And it's not a mind-related phenomenon, it's a real physical one. That's one of several unexpected twists and inversions that are found, when engaged in the rather pointless task of trying to pigeonhole the interpretation.

55. Unscrambling the omelette

As in the earlier quote from E T Jaynes, the omelette needs unscrambling. It's done in DQM by saying that the wave function (the omelette) contains both actual physical reality, and also statistical information. These are mixed up together, just as Jaynes said. Firstly, it's a real wave, which arises out of the superimposed set of all the possible positions for the flat dimensional axes when either light or matter travels along them, in a given situation.

But secondly, the wave function also contains statistical information, about a range of possibilities. These possibilities are for how this underlying action might be fitted into the relational level, which includes an emergent set of viewpoints. So the wave function describes a real wave of sorts, but it also doubles as a probability wave, which contains information about what might happen at the relational level.

And that's how the omelette is unscrambled. Both levels of reality are ontic, or real, in that they exist with or without observers. But the wave function is also epistemic, in that it contains information as well. I hope Jaynes would see this as a good answer to the conundrum he set out. And I also hope Matt Leifer will like this interpretation, having contributed a lot to the general attempts to solve the puzzle, by clarifying the different views of it. And for that matter, I hope William of Occam would have liked it, when he finally finished the 700 years of background reading he'd have to do.

56. A few final points

This book has not been about the 'big questions' on the universe, and in my view, nor has quantum mechanics. By that I mean questions such as how and why it got started, if it had a beginning. There are a lot of ideas and concepts that people think have bearing on these questions - I think more or less all of them have no bearing at all, and that if one is looking at the questions in a rational way, one should go to them without taking many assumptions along. And in my view one of the things on the list, of things that have no direct bearing on them, is quantum theory.

Incidentally, in this book occasionally the word 'man' is used, where some might prefer 'person'. It hardly needs saying, women have done great things in physics, both before and after the negative bias started getting removed - a few before and a great many afterwards. Most physicists now see men and women as equals, women are in leadership roles in physics, the old attitudes are fading fast. I've used the word 'man' simply where a turn of phrase is less funny or appropriate without it. It's like poetry, where sometimes changing a word would take away whatever was there. So in that kind of context, I've not gone by the blanket policing of language that some think necessary everywhere, regardless of context. Instead I've allowed the context to make a difference, as many others think it should.

There's a brief point to make about the scale issue in quantum mechanics, though I've mentioned it already. In DQM, the setup in the large-scale world is emergent. Something emergent is *needed*, because fundamental physical laws don't suddenly change at a given value for some parameter. So with or without DQM, there's a need for something emergent. That's hard to find, but the DQM picture fits that need.

But to build this emergent structure, one of the main premises is that lasting relationships between bits of matter get formed, through interactions. That might seem to require a lot of assumptions. And an explanation needs to explain as much as possible, using as few as possible (as in Chapter 24). DQM does well in this way, including here. The links it uses to build the framework don't require too many assumptions, because what we already know about entanglement, from both theory and experiment, strongly supports the idea that exactly that can happen - lasting relationships between bits of matter get formed - and that when it does, it happens via interactions.

As an interpretation, DQM might be seen as bringing in more assumptions, or 'adding stuff'. But something a bit like the dimensional setup I've used is widely believed to exist anyway, in Planck scale physics. So it doesn't need a lot of extra assumptions. Instead it just involves 'moving stuff around' - a picture that we've been building up in another area is brought into quantum mechanics, where it hasn't been used before.

It's worth making one last point, though it involves starting by restating an earlier one. Looking at the puzzle in a general way, we know that matter can be in something like a physical superposition. Apart from the wave function itself, there's not much else we know of that can do the same - that is, that can be in a physical superposition of alternatives. But the dimensions do that, and nowadays we tend to take them literally. They're set at every angle at once, if their positions are unspecified. That's a superposition.

So there you have a prime suspect for the cause of the quantum mystery, even before other reasons to suspect the dimensions appear, simply because nothing else does that. So to solve the puzzle, you bring in the only other thing we know of that does that. But it takes the DQM setup, sitting there underneath the picture, to make the role of the dimensions work.

Because the dimensions are set at many angles, if the angle (more accurately the orientation) can make all the difference, as in DQM, then each possibility in the wave function can arise from an angle. If so, that would allow this naturally existing superposition to be brought in, to help solve the puzzle of the other superposition - the one in quantum mechanics. So the question of whether or not the orientation can be relevant is vital, for instance if one is trying to assess DQM theory.

When I met Carlo Rovelli, and explained DQM theory to him, he said that it made sense to him less in terms of the dimensions, and more in terms of the orientation. (The conversation is in the documentary that was made on the subject.)

I won't go into the details of the conversation, but to me it's very important that there has been a positive response to the orientation aspect, from more than one physicist. Because the use of the orientation, as in DQM, *works* in quantum mechanics, for instance in RQM, that provides a crucial connecting link. It connects the two superpositions that I've mentioned above, and that connection is really a smoking gun - *it effectively reveals the answer*. That's because it makes the main point viable, which was about something that DQM does: it takes the only two physical superpositions that we know about in this world, and uses one to explain the other.

Anyway, I'll stop there, and hope that for some at least, this has helped to make sense of the nonsensical madness of quantum theory. I need to move on, and get to Book II as soon as possible, where rather than interpreting existing physics, I can set out some new physics, which will be of interest to some. However good the conceptual picture, some will have been wanting to get to grips with something that can be pinned down mathematically. And although I think conceptual physics has often been underestimated, I worked very hard to get to a testable theory, partly to make sure that the whole thing wasn't only an interpretation.

The background picture includes areas that can definitely be tested, both mathematically and in experiments. The first of these I've already tried at home, and it worked well - unfortunately, financial issues and the council's health and safety department won't let me try the second.

But there's a need to find out, in as clear a way as possible, if the landscape described here really is like the world at a small scale. And I'm glad to say that there will be ways to find out. To me this is a fascinating possibility, and Book II shows some principles that make testing the wider theory possible. Some of the predictions for experimental results are set out in mathematical detail, and for those who are interested, I'm hoping the experiments will be done, and then, well - we'll see what happens.

I'm also interested to see what happens once some of the conceptual ideas are no longer kept a secret, which my publisher and I have done until now. If you work for twenty years on something, you don't want it stolen when you finally try to publish a paper on it (it wouldn't be the first time that has happened). My publisher, Nigel Lesmoir-Gordon, has been amazingly helpful and supportive, and has stuck with me for seventeen of the twenty years. Recently, in 2017, we told the theory to two people in filmed conversations, but they agreed to keep it to themselves.

What I'm hoping is that there'll be increasing confirmation of the theory. It can be tested mathematically in many places. Incidentally, confirming it in that way is not solving the puzzle: if DQM is right, QM was a conceptual puzzle, not a mathematical one, and it could only be solved by conceptual thinking. But it's exciting that other physicists might start contributing to this, and we can of course do far better as a team than I can working alone.

And looking closely at DQM, people might find that quite a few things start clicking into place. This would be like when a large chunk of jigsaw is rotated through 90 degrees and placed differently, and it then turns out that whole edges join onto other edges in unexpected places. After a while the general crunching sound is heard of jigsaw pieces clicking into place in different areas, including in areas that had never even been thought about at all. This would be an exciting thing to happen, and because I believe the picture I've described is a good representation of the world that's out there, I think it'll fit neatly in many places. So it seems to me this will happen, and although the full background theory has more, I think this process will show what a conceptual picture can do. But we'll have to wait and see.

And standing a little further back, I'm also hoping that dimensional quantum mechanics will lead us along a path, and then further, then further along it again, taking us towards a new view of what's out there, and one that might help in better and deeper ways than just the basic ones. Beyond the purely theoretical side and the technology, the hope is that from this starting point there's a route that leads to a deeper understanding, and one that will open up wider horizons, and show us new things about the universe that none of us have even begun to think about yet, or to understand.

Additional note:

It took a long time for experiment to connect with the interpretation side of quantum theory, and to distinguish between the different pictures. In the 20th century it was more a matter of taste, as (apart from the verification of Bell's theorem), there wasn't much experimental evidence. But that's the vital place, and what we could do in the lab, and what we could do with the theory, got more and more sophisticated. And by 2018 people had found a way to test some ideas. What they found potentially supported some of the less mainstream views, such as Rovelli's relational interpretation.

An important result was published in February 2019, just before this book was due to come out, adding support to the view put forward here. A team in Edinburgh confirmed in the lab (via a similar setup) an idea published by a Zurich team in 2018, which shook up the quantum world, and which people have been talking about ever since. We naturally need both the idea and the confirmation, and in this case we now have both.

What the Frauchiger-Renner experiment shows, and what both teams found, is that the universe doesn't contain 'objective facts'. Different viewpoints can give different versions of the world. The result really does add to what we know. It means that one of three apparently commonsense assumptions has to be false. Without going into them at all here, this has bearing on several interpretations, each of which might need to let go of one assumption, and to find out which, and how. Rob Spekkens, at the Perimeter Institute, said it has helped to clarify some interpretations, as people have had to define their positions more exactly.

So it's not straightforward. But in the interpretations world, this strengthens the underdog, and weakens the front runners. The experimenters say: *"This result lends considerable strength to interpretations of quantum theory already set in an observer-dependent framework and demands for revision of those which are not."* So it's a major shake up. It potentially supports RQM, which predicts differences between viewpoints as one of its main elements. In RQM, it's absolutely to be expected that different viewpoints will show different versions of things - that's what RQM says happens. In an interview for New Scientist Magazine, Carlo Rovelli said he was delighted about it. And DQM, which is an interpretation of RQM, has exactly the same elements, and so is supported in the same way. Another less mainstream interpretation, QBism, is supported as well, because it has a subjective aspect.

So anyway, as our abilities with quantum mechanics improve, we start to get a handle on the very revealing interpretation side. An article and two papers on the discovery are to be found at the end of the reference section.

Index

Bohr, Nils	9-10, 55, 95
Conceptual physics	36, 48, 70-73
Copenhagen interpretation	9-10, 11
De Bono, Edward	41, 43
De Broglie, Louis	43
Decoherence	8, 16-18, 36, 61, 76-8, 84-6, 89
Dimensional quantum mechanics, DQM	59 onwards
Dimensions	58-9, 59-60, 67
Einstein	6, 39, 42, 50, 53, 71, 72, 79, 89
Emergence	24, 61, 69, 74-6, 84, 101, 102
Fields	79-80
Feynman, Richard	23, 64-5, 79
Frames of reference	33-4, 51-2, 68-9, 75, 76, 88-9, 91
Heisenberg, Werner	12, 55, 60
Information theory, quantum	30
Interactions	6, 16-19, 36-40, 56, 61-2, 69, 76-8 81-83, 86, 94
Interpretations	9-10, 11-13, 53, 60, 71, 74, 83, 100
Lateral thinking	41-43
Leifer, Matt	46, 48, 53, 101
Light (electromagnetic radiation)	14, 23-4, 39-40, 62-4, 79-81, 93-4
Many worlds interpretation	28, 86
Matter	8, 11, 14, 31, 33, 59, 62-3, 68, 75-8
Particle state	8, 10, 31, 39-40, 64
Photoelectric effect	39-40
Planck scale	67, 80, 102
Pusey, Barrett and Rudolph, PBR	12-13
Puzzle solving	42-48
Quantum eraser experiments	6, 13-15, 31, 95
Relational quantum mechanics, RQM	16, 33, 50-4, 60, 68-71, 77-8, 103
Renninger negative result experiment	94
Rovelli, Carlo	5, 16, 31, 51-2, 60, 69-71, 78, 103
Russell, Bertrand	68, 71
Schrödinger, Erwin	10, 60, 83-6, 95
Turok, Neil	5, 31-2, 57
Wave function	10-3, 18, 34, 54-5, 57, 76-7, 101-3
Wave-particle duality	24, 63-4
Wave state	31-2, 55, 64, 77, 86, 96
Wheeler, John	6, 24, 42, 50, 53, 64-5, 71, 73, 81
Zurek, Wojciech	18, 78

5	Introduction

Part 1. The core of the mystery
7

	1. Uncovering an unfixed world
8	2. An object made of possibilities
9	3. The cart and the horse
9	4. Interpretations
10	5. The wave function
11	6. What *is* the wave function?
13	7. Quantum eraser experiments

Part 2. The first signs of a change
16

	8. A new avenue of thought
17	9. Decoherence
18	10. Three stepping stones

Part 3. Ideas and clues
20

	11. Picking out the clues
21	12. The man who works in the laboratory - should he be removed?
23	13. The double slit experiment
25	14. The parts and the whole
26	15. Information and reality
27	16. Some different ways of seeing it
31	17. An 'exploratory' world?
32	18. A clue seen only with hindsight

Part 4. Interactions v. measurements
36

	19. Just the ones we know about
38	20. Why didn't we get to the idea of interactions sooner?
39	21. The photoelectric effect

Part 5. Finding new ideas
41

	22. Lateral jumps
44	23. Puzzle solving
45	24. Economy in assumptions
47	25. Narrowing it down

Part 6. Towards a solution
50

	26. A glimpse of something underneath the water

51	27. Light at the end of the tunnel?
53	28. We need an idea
54	29. Real and informational
55	30. Two questions

57	**Part 7. A possible explanation**
	31. The first question
58	32. The structure of the dimensions
59	33. Dimensional quantum mechanics
60	34. The second question
61	35. The laboratory
62	36. The wave-particle duality
65	37. Quantisation

67	**Part 8. Some other phenomena and concepts**
	38. The Planck scale
68	39. Two choices that were always seen as separate
69	40. A very direct property
70	41. Layers of interpretation
71	42. Conceptual physics
74	43. How real is a positioning of the dimensions?
76	44. Decoherence as in DQM
77	45. Another clue seen only with hindsight

79	**Part 9.**
	Using this picture to interpret what we observe
	46. Counterintuitive phenomena
80	47. Interpreting the double slit experiment
83	48. Schrödinger's cat
86	49. Entanglement
89	50. Instantaneous connections

93	**Part 10. Final points, and a little speculation**
	51. Removing some possible errors
95	52. What's inside an atom?
96	53. What else is inside an atom?
100	54. Defining this interpretation
101	55. Unscrambling the omelette
101	56. A few final points

References

References are listed in the format: Chapter number [page number].

Intro [6] **Einstein.** Conceptual basis, *Albert Einstein: Philosopher-Scientist*, from *The Library of Living Philosophers* Series, Cambridge University Press (1949).

Intro [6] **Wheeler.** John Wheeler made comments similar to this many times in his lectures and notes, according to one archivist more than a dozen times. But see 41 [69] *'It's not just a matter of nice simple formulas, there's some ideas out there waiting to be discovered.'* interview for 'Web of Stories', 20 seconds in. https://www.webofstories.com/play/john.wheeler/112

5 [11] **QM poll.** Maximilian Schlosshauer, Johannes Koer, Anton Zeilinger, *A Snapshot of Foundational Attitudes Toward Quantum Mechanics*, January 2013 arXiv:1301.1069v1, http://arxiv.org/pdf/1301.1069v1.pdf

6 [12] **Commentary on 1920s QM report.** Guido Bacciagaluppi, Antony Valentini, *Quantum Theory at the Crossroads: Reconsidering the 1927 Solvay Conference*, Cambridge University Press, October 2009 arXiv:quant-ph/0609184v2

6 [12] **Nature of wave function**. *Was Einstein Wrong?: A Quantum Threat to Special Relativity*, David Z Albert, Rivka Galchen, Scientific American Magazine, March 2009

6 [12] **PBR.** Matthew F. Pusey, Jonathan Barrett, Terry Rudolph, *On the reality of the quantum state,* 2011, http://arxiv.org/abs/1111.3328v3

6 [12] **Post PBR work, similar conclusion.** Martin Ringbauer, Ben Duffus, Cyril Branciard, Eric G. Cavalcanti, Andrew G. White, Alessandro Fedrizzi, *Measurements on the reality of the wavefunction*, 2014
http://arxiv.org/abs/1412.6213

6 [13] **Shan Gao,** *The Wave Function and Quantum Reality*, Proceedings of the International Conference on Advances in Quantum Theory, AIP Conference Proceedings 1327, 334-338 (2011). arXiv:1108.1187

7 [14] **Canary Islands.** Zeilinger et al, *Quantum teleportation over 143 kilometres using active feed-forward* Nature **489**, 269–273 (September 2012). http://arxiv.org/abs/1205.3909

8 [16] **RQM.** Rovelli, C., *"Relational Quantum Mechanics"*; International Journal of Theoretical Physics **35**; 1996: 1637-1678; arXiv:quant-ph/9609002

9 [17] **Decoherence**. Schlosshauer, M. 2005, *Decoherence, the measurement problem, and interpretations of quantum mechanics*. Reviews of Modern Physics, 76 1267-1305, http://arxiv.org/abs/quant-ph/0312059

9 [18] **Zurek**, W. H., 1991, Phys. Today 44, 36. Updated: quant-ph/0306072.

9 [18] **Decoherence confirmed by experiment.** Chiorescu, I., Nakamura, Y., Harmans, C. J. P. M. & Mooij, J. E. *Coherent Quantum Dynamics of a Superconducting Flux Qubit,* Science **21**, 1869–1871 (2003).

14 [26] **Bohm**, *Wholeness and the implicate order*, David Bohm, Routledge 1980. ISBN 0-203-99515-5

17 [31] **Jan Westerhoff**, *Reality: is matter real?*, New Scientist Magazine, 26th Sept. 2012 https://www.newscientist.com/article/mg21528840-700-reality-is-matter-real

22 [43] **de Bono**, *The five day course in thinking*, Edward de Bono, first published by Signet Books, November 1968

25 [47] **QM complete**. Roger Colbeck; Renato Renner (2011). *No extension of quantum theory can have improved predictive power*.
Nature Communications **2** (8). arXiv:1005.5173.
http://www.nature.com/ncomms/journal/v2/n8/full/ncomms1416.html

28 [53] **Leifer**, M, *More on criteria for interpretations*
https://mattleifer.wordpress.com/2006/07/06/more-on-criteria-for-interpretations

28 [53] **Wheeler.** Rederiving quantum theory. Transcript of a 1980s BBC radio interview, *The Ghost in the Atom: A Discussion of the Mysteries of Quantum Physics*. J. R. Brown [edited by], P. C. W. Davies, Cambridge University Press, 1986. ISBN: 9780521307901

28 [54] **Fuchs**, Christopher, *Quantum mechanics as quantum information (and only a little more)*, in A. Khrenikov (ed.) *Quantum Theory: Reconstruction of Foundations* (Växjö: Växjö University Press, 2002).
http://arxiv.org/abs/quant-ph/0205039

29 [55] **Jaynes**, E.T. *Complexity, Entropy and the Physics of Information* (ed. Zurek, W.H.) 381 (Addison-Wesley, 1990).

41 [70] **Kochen**, S. Symposium of the Foundations of Modern Physics: 50 Years of the Einstein-Podolsky-Rosen Gedanken experiment (World Scientific Publishing Co., Singapore, 1985), pp. 151–69.

42 [71] **Wheeler.** 'It's not just a matter of nice simple formulas, there's some ideas out there waiting to be discovered.' interview for 'Web of Stories', 20 seconds in. https://www.webofstories.com/play/john.wheeler/112

46 [79] **Einstein**, Albert (1924) 'Über den Äther', *Verhandlungen der Schweizerischen Naturforschenden Gesellschaft* 105:2, 85-93. Also published in English: S.W. Saunders, translator. *The Philosophy of Vacuum*, edited by Simon Saunders and Harvey R. Brown, pp. 13-20; Clarendon Press, Oxford, 1991. ISBN 0-19-824449-5

46 [79] **Gamow.** George Gamow, *One, Two, Three... Infinity* (1947, revised 1961), Viking Press, reprinted by Dover Publications, ISBN 978-0-486-25664-1

46 [79] **Feynman.** Richard P. Feynman, Robert B. Leighton, and Matthew Sands, The Feynman Lectures on Physics, Addison-Wesley, Reading, MA, Vol. 1, 1964, p. 2-4.

48 [84] **Experimental evidence for decoherence.** J. F. Poyatos, J. I. Cirac, P. Zoller: *Quantum reservoir engineering with laser cooled trapped ions.* Phys. Rev. Lett. 77, 4728–4731 (1996).

51 [93] **Realism,** Travis Norsen, *Against 'Realism'*, Foundations of Physics, Vol. 37 No. 3, 311-340 (March 2007)   arXiv:quant-ph/0607057v2

51 [94] **Renninger.** M. Renninger, (1953) Zeitschrift für Physik, **136** p 251

51 [95] **Angelo**, R. M., *On the interpretative essence of the term "interaction-free measurement": The role of entanglement*   arXiv:0802.3853v3

54 [100] **Fuchs**, Christopher, http://perimeterinstitute.ca/personal/cfuchs

Additional note [106] **Frauchiger-Renner experiment.** Article:

https://www.newscientist.com/article/2194747-quantum-experiment-suggests-there-really-are-alternative-facts

Frauchiger, D., Renner, R., *Quantum theory cannot consistently describe the use of itself*, Nature Communications 9(1), 3711 (2018), arXiv:1604.07422v2

Proietti, M., Pickston, A., Graffitti, F., Barrow, P., Kundys, D., Branciard, C., Ringbauer, M., Fedrizzi, A., *Experimental rejection of observer-independence in the quantum world,* arXiv:1902.05080 (Submitted 13 Feb 2019).

Lightning Source UK Ltd.
Milton Keynes UK
UKHW020638230120
357484UK00005B/717

9 780956 422262